# FOOD PRESERVATION AND IRRADIATION

# FOOD PRESERVATION AND IRRADIATION

**S.N. Mahindru**
*Chief Consultant,*
Food Standards and Laws,
MDH, Ltd. New Delhi

**A P H PUBLISHING CORPORATION**
**ANSARI ROAD, DARYA GANJ**
**NEW DELHI-110 002**

*Published by*
S.B. Nangia
**A P H Publishing Corporation**
4435-36/7, Ansari Road, Daryaganj
New Delhi 110002
Ph.: 23274050
E-mail : aphbooks@gmail.com

2023



*Printed at*
**Balaji Offset**
Navin Shahdara, Delhi 110032

# PREFACE

Microorganisms, parasites and insects are ambiquitos in nature. Some of them are pathogenic and the rest non-pathogenic. Even the very best food manufacturing processes, based on most careful attention to hygiene cannot completely rid the processed foods from microorganisms.

Food irradiation is a physical process for the treatment of foods akin to conventional process like heating or freezing. It prevents food poisoning (salmonellosis), reduces wastage to contamination and at the same time preserves quality.

With the exception of irradiation of spices and dried vegetable substances, which is wide spread, other applications of food irradiation technology remain marginal. Misconceptions about whether irradiation food is safe to eat and how irradiation can complement or replace other methods of food preservation are largely responsible for the above situation. Besides the beneficial results of food irradiation are not generally available to individual consumers, families and societies.

Today enormous food losses is one of the main problems confronting developing countries, tropical countries in particular, like India, where insect infestation and microbial ingress confront their post harvest agriculture and products. Through timely irradiation, these unwarranted losses can be greatly reduced. Also the shelf-life of food items can be extended leading to export of the food products on a more sustained basis.

Lingering doubts about the safety of irradiation foods have been laid to rest by world renowned Indian scientists like Drs. R. A. Mashelkar and M.G.K. Menon, both FRS.

# CONTENTS

Chapter 1

# INTRODUCTION

Micro-organisms, parasites and insects are umbiquitos in nature. Some of them are pathogenic and the rest nonpathogenic. Even the very best food manufacturing processes, based on most careful attention to hygiene cannot completely rid the processed foods from microorganisms. Food irradiation has the same objectives as those of the more traditional type of food preservation processes.

(i) The prevention of Salmonellosis i.e. food poisoning.

(ii) The reduction of wastage due to contamination.

(iii) Preservation of quality.

Food irradiation is a physical process for the treatment of foods akin to conventional processes like heating or freezing. In this case food is exposed to gamma ($\gamma$) rays kept inside a shielded structure either packed or in bulk, for a specific time to deliver the desired dose. The sources of radiation used are of two types, isotope sources like $^{60}$Cobalt ($^{60}$Co) or $^{137}$Caesium ($^{137}$Cs) emitting $\gamma$-rays, or X-rays and electrons generated by machine sources. Gamma and X-rays are elctro-magnetic radiations of very short wave length of same nature as ultraviolet (UV), visible infrared (IR), microwave etc. Specific quantity of ionizing radiation absorbed (Gray (Gy) by the food and the maximum limit of strength of sources used are 10Mev (Milli electron Volt) for electron, 5 MeV for X-rays and $^{60}$Cobalt of 1.33 MeV and $^{137}$Caesium of 0.66 MeV. At these energy levels no induced radio activity can occur in food elements. Even a dose 100 time higher than the dose prescribed for food irradiation does not introduce radio activity.

In other words when a patient is X-rayed, visible rays penetrate his body and throw a shadow on a film placed behind or underneath his body. These rays consist of infinitesimally minute particles of electrically charged matter and they vibrate in a wide scatter of wave lengths. Gamma($\gamma$) radiation is similar to X-radiation but the wave lengths are contained within a narrow band; X-rays-$10^{-9}$ ~ $10^{11}$; Gamma($\gamma$) rays $10^{11}$ ~ $10^{13}$ (meter); it has similar powers of penetration. Just as the rays penetrate the body so they can penetrate the foodstuffs, and the principal behind radiation of food is to use the power of the rays to penetrate the food, while adjusting the dose of the rays so that germs are destroyed but the food left is not damaged in any way.

The treatment of food with ionizing radiation for its preservation is a resurging technology which was lying dormant since its introduction after the discovery on the use of X-rays as an effective way to kill bacteria in food. A patent for this purpose was taken in 1905 in U.K. but the X-rays machines available during that period were not suitable for irradiation of food as we perceive to day. The late forties saw the emergence of other gamma sources which enthused the research on $\gamma$-irradiation. Large research programs were initiated on the use of $\gamma$-rays for food preservation in many countries including USA, U.K. and Soviet Union. After seeing the promising result world wide, research programmes began under the aegis of International Atomic Energy Agency now the well know (IAEA) and more than 55 countries conducted strong R & D work on food irradiation. The detailed history of wholesomeness determination of irradiated foods follows in the in the next chapter.

According to Dr. F.S. Antezana, Deputy Director General *(ad interim)* three Organisations had a long and successful history of collaboration in the area of food irradiation, which had started as early as 1961. In 1980, the Joint Expert Committee on the wholesomeness of irradiated food had concluded that "......irradiation of any food commodity up to an overall average dose of 10 kGy presents no toxicological hazard......... and introduces no special nutritional or microbiological problems". (WHO Technical Repeat Series, No. 659.)

These conclusions clearly established the wholesomeness of any food irradiated up to an over all average dose of 10 kGy. The reasons for this limitation were essentially two fold.

(i) The 1980 Joint Expert Committee was asked to assess the wholesomeness of irradiated foods on the basis of data available at that time, which mainly concerned doses below 10 kGy.

(ii) Many of the anticipated applications for irradiation of food would require doses of less than 10 kGy. Examples of such applications include:

(a) The elimination of vegetative bacterial pathogens from foods like meat, poultry, fish, and fresh fruits and vegetables.
(b) The inhibition of sprouting in potatoes and other tubers.
(c) The insect disinfection of grains and dried fruits such as dates and figs.
(d) Extension of the shelf-life of refrigerated foods and
(e) The treatment for quarantine purposes of fruits and vegetables.

On the basis of the scientific judgment provided by the Joint Expert Committee in 1980, as well as additional supportive evidence, the FAO/WHO Codex Alimentarius Commission adopted in 1983, the Codex General Standard for Irradiated Foods, limiting the overall average dose to 10 kGy. As a consequence, a large number of governments (currently 40) initiated regulatory actions permitting the irradiation of a considerable number of food commodities.

With the exception of irradiation of spices and dried vegetable substances, which is wide spread, other applications of this technology remain marginal. Misconceptions about whether irradiated food is safe to eat and about how irradiation can complement or replace other methods of food preservation are largely responsible for this situation. Consequently, the beneficial results of food irradiation–the improvement of hygienic quality of certain foods and the reduction of post-harvest losses are not generally available to individual consumers, families and societies. There are indications, however, that irradiation will be increasingly used to ensure hygienic quality of food of animal origin and to overcome quarantine barriers in trade in fresh fruits and vegetables. Moreover, the use of high-dose irradiation could also result in less dependence on refrigeration of food, which is an energy-intensive technology.

The fact that the international organizations and the Codex limited the dose level to 10 kGy has frequently been interpreted as meaning that this is a dose above which toxic substances could be introduced or nutritional adequacy of foods could be negatively influenced. However, there are current applications of food irradiation involving doses above 10 kGy which indicate that this is not the case. These include the development of high-quality shelf-stable convenience foods for general use and for specific target groups, such as immunosuppressed individuals and those under medical care. Such shelf-stable foods have also been used successfully by astronauts, military personnel and outdoor enthusiasts in some countries.

The presence in food of pathogenic microorganisms, such as *Salmonella* species, *Escheria coli* 0157:H7, *Listeria* & *monocytogenes* or *Yersinia enterocolitica*, is a problem of growing concern to public health authorities all over the world. In an attempt to reduce or eliminate the resulting risks, measures such as strict hazard analysis critical control paint (HACCP) system regulations have been issued in many countries. To ensure that processed raw meat and poultry products are free of contamination and are consistently free of pathogens, irradiation to a dose of 10 kGy or less could be considered Critical Control Point in the HACCP plan for there products. In some instances, however, an upper limit of 10 kGy for the over all average dose could preclude the effective use of this method.

In the case of spices irradiation this need for a higher average dose has already been recognized in several countries. France permits an average dose of 11 kGy, India 10 kGy for the irradiation of spices and dry aromatic substances, Argentina and USA permit a maximum dose of 30 kGy ($D_{max}$) for this purpose.

In India, extensive R&D work conducted at Food Technology and Enzymes Engineering Division, BARC, has established the process parameters for using this technology for the preservation of a number of foods and food products which have immense economic importance. The main objectives are:

(i) Prevention of post-harvest food losses.

(ii) Ensuring hygienic qualities of food.

But there is a criteria of using food irradiation. Following are the critical aspects considered while any agricultural or food and its products are subjected to radiation processing. Of radiation-processed foods include investigations on:

(a) Wholesomeness of the irradiated product for consumption.

(b) Possibility of included radioactivity .

(c) Microbiological safety.

(d) Safety of chemical changes.

(e) Nutritional adequacy.

## POST HARVEST LOSSES

The post harvest practices including inadequate storage and preservation facilities, as well as adverse climatic conditions, cause heavy losses in Indian agricultural and marine produce. Food worth thousands of crores is lost due to insect infestation, microbial contamination, physiological changes like sprouting etc. These have been summarily presented below :–

### (a) Sprouting inhibition of tubers and bulbs

Sprouting losses in potatoes and onions are very large since enough storage facilities are not available to cope with enhanced production of these commodities. In the case of potatoes, cold storage (2-4$^0$C) is practised for shelf life. But after irradiation, the crop can be stored at 15$^0$C with minimum rotting losses. This saves not only energy and storage cost but also improves eating and processing qualities of there potatoes.

Irradiated onions can be stored under ambient condition with forced ventilation for more than six months.

Irradiation has shown success in preservation of yam also.

### (b) Shelf-life extension

Low dose of irradiation upto 1 kGy can interfere with physiological process of senescence of certain fruits like mango, banana, papayas and vegetables like mushroom and asparagus. By combining radiation with hot water dip (50°C for 5 min) both delay ripening and disease control of certain fruits like mango and papayas. The dose levels required for extension of shelf life in many cases also help to kill eggs and larvae of fruit fly and seed weevil. Shelf life of strawberries can be extended at a dose of 2.5 kGy.

### (b) Disinfestation of stored food products

Cereals and pulses, dried fruits and nuts, dried fish and meat face quality deterioration due to infestation by insects. Fumigation of dried foods and ingredients with various chemicals like ethylene dibromide (EDB), methyl bromide (MB) and phosphine ($PH_3$) was the prevalent method in many countries for control of pests. Now these chemicals are being banned in many countries due to their potential health hazards. Many food processing industries are looking upon irradiation as a viable alternative to fumigation. Irradiation at 0.5 ± 0.2 kGy dose, done under suitably packed samples to prevent reinfestation eliminates insect problem in dried fruits and nuts, wheat products, pulses, dried fish, etc. Unlike chemical fumigation, irradiation does not leave any residue. A growing concern about fumigation is that insects develop resistance to chemicals but irradiation does not produce any radiation resistant strain of insect.

## ENSURING HYGIENIC QUALITY OF FOOD

### (i) Microbial hygienic quality of food

Radurization process is capable of eliminating the spoilage microorganisms from fish, meat, fruits, vegetables etc. and all the spoilers are sensitive to low dose of irradiation. A dose of 3 ± 2 kGy is satisfactory to achieve this goal. Several fold reduction of spoilage organism can extend shelf life to 4-fold under identical storage conditions. But irradiation is not a substitute of good manufacturing practices (GMP) and the product brought for irradiation should be of optimal quality in terms of freshness and maturity.

In processed foods spices and other dried ingradients are increasingly used because of the demand for convenience foods like canned meat, sausage, salami, soup mixes etc. The microorganisms present in spices may grow well and contaminate the product. This is the reason that importing countries control the microbiological quality of spices. With the regulatory restrictions on the use of chemicals in food, irradiation at 10 kGy is the only alternative for decontamination of spices and condiments.

### (ii) Elimination of food borne pathogens

Food borne diseases are in an increasing trend and the situation is alike in both developing and developed countries. Food from animal origin is contaminated by pathogenetic microorganisms and parasites such as *Salmonella*, *Compylobacter*, *Listeria*, *Toxoplasma, Trichinella*, *Ascaries*, and *Fascoila* etc. Salmonella contamination leads to rejection of the consignment. A medium dose of radiation 4 ± 3 kGy eliminates Salmonella and other pathogens even from frozen prepacked samples, which no other technology can achieve.

γ-irradiation has certain advantages over other methods. It has very great powers of penetration and can be used to sterilize food after it has been canned or packeted. It can penetrate large joints or even carcasses, and bulked quantities of goods such as feeding stuffs. It produces scarcely any heat and can be used on frozen foods such as poultry or frozen meat without raising the temperature more than a degree or two, if at all: in fact, radiation is possibly more effective at very low temperature and the dose can be reduced accordingly. It may be used to sterilize food completely, but sometimes a smaller dose is used to achieve only a reduction of germs.

The term pasteurization is used for this purpose just as it was in the canning of ham but the difference with the radiation pasteurization is complete, whereas with heat pasteurization of solid food penetration cannot be guaranteed.

The irradiation dosage used for food stuffs probably acts mainly by energizing the atoms in molecules of water in the food and these, the Ḣ & OḢ radicals, then become highly active chemically as reducing and oxidising agents. Direct action on organic matter is minimal.

Irradiation is meant for destroying spoilage germs so as to prolong the keeping life of food; also to destroy disease germs so as to make it safe to eat. Obviously irradiation can be used to sterilizes animal food stuffs. This would attack salmonellas infection at one of its main sources.

Irradiation would destroy moulds on fruits hence make transport and distribution easier. It has been already allowed for mangoes, raisins, figs and dried dates only. Under PFA Rule 74 (6 to 8) but not for other fruits.

## Present Status

Dr. Anil Kakodkar, Chairman, Atomic Energy Commission feels that the potential of radiation technology is still under utilized in the area of food processing though it holds a great promise. Awareness amongst the masses about the safe use of radiation processed foods is thus of paramount importance.

*Message on National Worship on Awareness of Gamma Radiation Processing of Foods, Spices, Animal Feed and Health Care Products, Nov 28-29, 2003.*

According to B. Bhattarcharjee, Director, Bhabha Atomic research Centre, Bombay, there are more than 200 commercial radiation plants distributed world wide today, which carry out sterilization of a variety of medical and food products. Recently, the scientific community has realized the enormous potential of ionizing radiation to hygienize food and agricultural products, thereby extending their shelf life. *loc. cit*

According to Shri R.C.A. Jain, Secretary, in the Ministry of Agriculture, Government of India :-

"Gamma radiation processing of food, spices, animal feed and health care products, has assumed great significance for substantially reducing food losses, which is one of the main problems confronting the developing countries to day. India being a tropical country, insect infestation and microbial ingress are problems confronting our post harvest agriculture produce and products. Through the timely irradiation, unwarranted losses can greatly be reduced as well as enhance shelf-life of the food items besides the export of the products on a more sustained basis" *loc. cit*

Dr. S.P. Agarwal, Director General Health Services, Government of India, admits that the technique of radiation which is non-additive and cold process that does not destroy heat labile constituents, may prove to be of significant value for enhancement of shelf life of agro-products in their natural form as well as for the safety of health care products. Although the methodology involved is eco-friendly and safe to workers, still there is considerable lack of awareness about this technique and people have doubts about the safety of irradiated foods.

It is queer yet paradoxical about the above lingering doubts about the safety of irradiated foods despite world famous Indian scientists like Drs. R.A. Mashelkar and MGK Menon, both F.R.S, state categorically that

"Radiation processing of food items and sterilizing of medical and healthcare products, is widely accepted all over the world. This technology has great significance for reducing wastage and enhancing safety of food items and of health care products."

**Chapter 2**

# HISTORY OF WHOLESOMENESS DETERMINATION OF IRRADIATED FOODS

Extensive animal feeding studies designed to detect any toxic factors that might be present in various irradiated foods were carried out in 1950's and 1960's, mostly in the United Kingdom (U.K) and the United States (U.S.A). On the basis of these studies a Working Party was established. U.K. Ministry of Health agreed that extensive tests on a wide range of foods, carried out particularly in the U.S.A., have yielded no evidence of the formation of carcinogens in irradiated food. The Working Party, after considering the effects of irradiation on nutrients, on the possible presence of induced radioactivity, and on possible microbiological hazards of irradiated food, also concluded.

"That the evidence for the wholesomeness of food which has been irradiated under specified and closely controlled conditions is reassuring" - *U.K. Ministry of Health (MOH)*, HMSO, 1964.

The U.S. Army Surgeon General concluded that,

"Foods irradiated upto an absorbed dose of 5.6 Mrad (56 kGy) with a $^{60}$Co source of radiation or with electrons with energies upto 10 million electron Volts (MeV) have been found to be wholesome, i.e., safe and nutritionally adequate" –

*–Radiation processing of foods. Hearing before the Congress of U.S.*, 9th and 10th June 1965.

At about that time, however, the United States Food and Drug Administration (USFDA) and other national health agencies began to apply more stringent criteria for safety testing. The above reported animal feeding studies (1950) were deemed inadequate. In response, a massive programme of re-evaluation of safety assessment of radiation-sterilized beef as well as chicken meat was initiated.

FAO, IAEA and WHO held the first international meeting in October 1961, to discuss the wholesomeness data and legislative aspects of irradiated foods. Participants from 28 countries attended it. The presented data was not deemed adequate hence the meeting decided that general authorization of the commercial use of radiation for the treatment of food would be premature. Instead the above three should consider an early setting up of a Joint Expert Committee to advise on the special requirements for the testing of the wholesomeness of irradiated foods.

The above referred Joint FAO/WHO/IAEA Expert Committee (JEC) on the technical basis for legislation on irradiated food met in Rome during April 1964. Observations of the JEC were:

"Extensive tests conducted by feeding to animals and to a lesser extent to human volunteers irradiated food treated in accordance with procedures that should be followed in approved practice have given no indication of adverse effect of any kind and there has been no evidence that the nutritional value of irradiated food is affected in any *important way*"

–(1966) (*WHO Technical Report Series, No. 316*).

The JEC recommended legal control of irradiated food "by the use of a list of permitted foods irradiated under specific conditions" and made recommendations as to which tests should be applied to an irradiated food to establish its safety for consumption; it suggested that these tests should be broadly similar to those used for testing the safety of food additives. When the JEC met in Geneva in April 1969 i.e., after complete five years, it gave *temporary acceptance* to irradiated potatoes (doses upto 0.15 kGy) and to wheat and wheat products (doses upto 0.75 kGy), but found the available data on irradiated onions to be unsatisfactory for an evaluation.

–(1970) (*WHO Tech. Report Series, No. 451.*)

The acceptance for potatoes and wheat was designated as *temporary* because the available data were insufficient to fully establish safety; additional evidence within a specified period of time was summoned.

Consequently, the International Project in the field of food irradiation was created in 1970. Under the sponsorship of FAO/ IAEA/OECD (Organisation for Economic Cooperation and Development) 24 counties jointly began addressing the related issues. Feeding studies contracted by the International Project were carried out with irradiated wheat flour, potatoes, rice, iced ocean fish, mangoes, spices, dried dates and cocoa powder. WHO became advisor. The International Project limited its studies to the dose range upto 10 kGy. By 1982, this project completed its assigned studies and was consequently wound up. The JEC gave an unconditional acceptance to wheat (upto 1 kGy) potatoes (upto 0.15 kGy), papayas (upto 1 kGy); strawberries (upto 3 kGy), chicken (upto 2 kGy), while onions (upto 0.15 kGy), rice (upto 1 kGy), and fresh cod and red fish (upto 2.2 kGy) received provisional acceptance i.e., temporary acceptance because in these foods some additional testing was deemed necessary.

Irradiated mushrooms needed further assessment hence not cleared (1977). *WHO Tech. Series, No. 604.*

The Committee clearly defined that irradiation is a physical process of treating foods hence is comparable to other methods of food preservation like heat and freezing. It also recognized the value of chemical studies as basic for evaluating the wholesomeness of irradiated foods.

In its October, 1980 meeting in Geneva, the JEC, on the basis of huge additional data provided by the aforesaid International Project drew the following conclusions.

(a) None of the toxicological studies carried out on a large number of individual foods, representing different classes of food having similar chemical composition, had produced evidence of adverse effects as an after effect of irradiation.

(b) Supporting evidence was provided by the absence of any adverse effects resulting form the feeding of irradiated diets to laboratory animals, the use of irradiated feeds in livestock production and the practice of maintaining immunologically incompetent patients on irradiated diets.

(c) Radiation chemistry studies had shown that the radiolytic products of major food components were identical, regardless of the food from which they were derived. Moreover, for major food components, most of these radiolytic products had also been identified in foods subjected to other accepted types of food processing. Knowledge of the nature and concentration of these radiolytic products indicated that these was no evidence of a toxicological hazard.

The committee therefore concluded that irradiation of any food commodity upto an overall average dose of 10 kGy presented no toxicological hazard, hence, toxicological testing of foods so treated was no longer required.

The Committee further concluded that the irradiation of food upto an overall average dose of 10 kGy introduced no nutritional or microbiological problems. However, it emphasized that attention should be given to the significance of any changes in a particular irradiated food in relation to its role in the diet.

The Committee recognized that higher doses of radiation were needed for the treatment of certain foods, but considered that the available data was insufficient for a toxicological evaluation and wholesomeness assessment of foods treated and that further studies in this area were needed. The final results of the studies carried out in the U.S.A. on high-dose irradiated food items were not available at that time.

An *ad hoc* group of experts, invited by WHO, reviewed and evaluated scientific studies conducted after 1980 JEC meeting, including studies on the high-dose (59 kGy) irradiation of chicken carried out in U.S.A. as well as many of the older studies that had already been considered previously. The report of this evaluation was published by WHO. The group calculated that irradiated food produced under established good manufacturing practices (GMP) could be considered safe and multinationally adequate because the process of irradiation would:-

(a) not lead to changes in the composition of the food that, from toxicological point of view, would have an adverse effect on human health;
(b) not lead to changes in the microflora of the food that would increase the microbiological risk to the consumer;
(c) not lead to nutrient losses to an extent that would have an adverse effect on the nutritional status of individuals or populations.

The International Consultative Group on Food Irradiation (ICGFI) decided in 1989 to assemble, with the help of consultants and in collaboration with WHO, all relevant data on radiation application involving doses above 10 kGy to determine whether or not the information available would be adequate for an assessment of the wholesomeness of food irradiated to these doses. On the basis of reports written by several experts, a consultation was held in 1994 in Karlsruhe which concluded that the available data on radiation chemistry, toxicology, microbiology and nutritional properties of food were adequate for this purpose. *(unpublished document, WHO/FNUY, FOS/95.10.)*

Loss of vitamins is often referred as an argument against food irradiation. It is true that some vitamins are more sensitive to irradiation. But the losses occurring when whole food is irradiated at the recommended dose is much less than those observed with studies on pure vitamin in solution using different doses. Vitamins losses occur in many other processes like drying, canning, smoking etc. and these processes are still used and no body asks for banning them–argued PM Nair of Food Technology & Enzyme Engineering Division, BARC, Bombay in a symposium on 16-18 March, 1994.

Consequently by April 1993, several countries had irradiation facilities available for food processing.

In India Government of India after evaluating all the data available, in 1986 approved in principle the adoption of food irradiation, as a method of preservation of food stuffs subject to an over all average doze of 10 kGy. A National Monitoring Agency (NMA) was constituted under the Ministry of Health & Family Welfare to oversee all aspects regarding the implementation of this technology. The composition as well as functions of various Ministries in promoting this technology have been spelt out by Ministry of Health and Family Welfare. NMA, at the very first instance had cleared irradiated items like spices, onions and frozen sea foods for export and subsequently spices, onions and potatoes for domestic consumption. In order to control the irradiation of food based on Codex General Standards, Atomic Energy (Control of Irradiation of Food) Rules 1991 were published in Gazette of India, March 1991. For trade in irradiated food, NMA had approved the amendment to introduce food irradiation under Prevention of Food Adulteration Act & its Rules (PFA) and these revised amendments and final publication appeared as Part XVII - Irradiation of Food, Rules 73 to 78 vide GSR 614(E), dt. 9.8.1994, corrected by GSR 63(E) dt. 7.2.1995.

Rule 73 details meanings of various technical terms used in this chapter, like "irradiation," "irradiation facility," "operator of radiation faculty," types of irradiation like x-rays, $\gamma$-rays and electrons.

Rule 74 specifies the names of foods permitted for irradiation and the dose of irradiation fixed for each food, minimum, maximum and overall average. To begin with, as stated earlier, only onions, spices and potatoes were allowed. In 1998, rice, somolina *(sooji)* or *rawa*, wheat atta and maida, mangoes, raisins, figs, dried dates, ginger, garlic and shallots (small onions), meat and meat products including chicken were allowed vide GST 172 (E) dt. 6.4.1998. Fresh sea foods, frozen sea foods, dried sea foods and pulses were allowed vide GSR 320(E) dt. 2.5.2001.

Rule 75 details requirements for the process of irradiation, Rule 76 puts restrictions on the dose limit and the radiation source as laid down under the Atomic Energy (Control of Irradiation of Foods) Rules, 1991. It also requires to put identification mark for being an irradiated food lest it may be again subjected for reirradiation. Only specifically permitted foods can be re-irradiated.

Rule 77 is all about maintaining records of irradiation of food.

Rule 78 clearly enunciates that the irradiated foods shall comply with all the provisions of PFA Act 1954, and rules made there under specifying standards of such food.

Introduction of any new technology always faces apprehensive attitude. Many people are technology averse in general and more so when food is concerned. Radiation in the context of food faces stiff opposition. Generally irradiation is confused with radioactivity though this is not so in the context of food. Irradiated food may become radioactive is not a scientific truth, yet there is an active propaganda against food irradiation form anti-nuclear movement. One of their objections is

"Safety had not been absolutely proven."

Factually no body can prove absolute safety. However, long animal feeding studies may be carried out, the critic can still say "Man is not a rat, so your experiments prove nothing".

PM Nair of BARC had long back suggested that to gain consumer's confidence the only way is to disseminate factual information, via different media, and demonstrate the efficacy and benefits of the food irradiation technology to food industry, consumers, including different target groups. Market demonstrations of irradiated foods where consumers themselves can see the advantages in terms of safety and cost benefit factor will go a long way to booster the acceptance of these foods. It is well known that market testing and commercial sales of irradiated foods conducted by more than 20 countries were all positively in favor of food irradiation. None of these studies showed that consumers would reject irradiated foods. A close cooperative effort between food companies, irradiation facilities and retailers shall certainly be helpful for the success of this mission.

Notwithstanding, the future of food irradiation industry still hinges on two important factors

(i) Government's authorization and labelling regulations.
(ii) Monetary benefit to the food processors.

Food irradiation cannot be done unless the concerned government permits it. The governments, despite fully aware of the advantages of food irradiation vis a vis other processing technologies, are under political pressures exerted by those who oppose this technology for their political advantage rather than technical merits. Labelling is another obstacle in this case as shown below:—

(a) Under Rule 42 (ZZZ)(7), all packages of irradiated food shall bear the following declaration and logo, namely:
Processed By Irradiation Method
Date of Irradiation.... ( o )
Licence No.... ( ∞ )
Purpose of Irradiation....

(b) Under Rule 44F, there is the restriction, that irradiation food shall be offered for sale only in prepackaged conditions.
(c) Under Rule 48D, *Storage and Sale of irradiated food* i.e. save as otherwise provided in these rules no person shall irradiate for sale, store for sale, or transport for sale irradiated food.
(d) Under Rule 49 (26) *Conditions for sale of irradiated food* i.e. all irradiated foods shall be sold in prepacked condition only. The type of packaging material used for irradiated food for sale or for stock for sale or for exhibition for sale or for storage or sale shall conform to the requirements of packaging material as per Rule 49; sub Rule 1 to 5(vi).

In human society, rules are framed to rectify the prevalent defects in a system but result in harming the society.

"As of today agricultural products, foods in particular, are handled in very rough manners, thereby causing irreversible physical damage. If such state of affairs continues after irradiation of food, we will be frittering away the efforts undertaken during processing. This could be minimized to a great extent if we concurrently improve over packaging systems and materials and attempt to pack only say 15-20 kgs/pack, preferably using card board and plastic films. The finished product storage should be adequate as there is a tendency by the customers to utilize the space after irradiation as a rent free godown. Strict management discipline and panel rates need be imposed on defaulters."

Source– J.P. Kapur, "*Welcome Address, National Seminar on Application of Radioisotopes and Radiation Technology in Food Processing and Health Case Products.*" Sept. 22, 1999, New Delhi.

Agreed that the laws regarding food irradiation in India are too severe.

"Laws too gentle are seldom obeyed; too severe, seldom executed'- Franklin - Poor Richard's Almanac.

We should take a hint form the words from "*Opinion*" -

"The law is not an end in itself, nor does it provide ends. It is preeminently a means to serve what we think is right" - W.J. Brennan, Jr. in "*Opinion.*"

At the same time, there is a certain psychological barrier in the minds of the consumers about long term effect on human biological system affecting chromosomes about consuming irradiated food as has been the recent opposition to genetically modified soyabean and corn in Europe. Although scientifically there is reported evidence that no mutagenesus adverse changes take place due to long term consumption or radiolytic and chemical changes during irradiation.

We know that more than 30 countries regularly consume products like spices, fish, potatoes, onions, fruits like apple, mangoes etc.

The issues therefore pertain to public awareness of facts and transparency of scientific data. These facts should detract apprehensions of the consumers. Though labelling of irradiation food is quite important and desirable for transparent operation, but it should not be mandatory as stated earlier. In U.S.A. labelling is optional and not mandatory. We should replace the irritant label of "Irradiation" with "Ionically pasteurized/sterilized."

## Wholesomeness of Irradiated Foods

Five aspects of wholesomeness must be considered, the first two are unique to irradiation, the other three apply equally well to thermal sterilization:-

### *(i)Induced Radioactivity*

The induction of radioactivity in food depends on

(a) the type of irradiation and its energy,
(b) the dose applied to the food,
(c) the abundance of specific elements in the food,
(d) the half life of the induced radioisotope is important.

If the induced isotopes have a short half-lives than these isotopes disappear before reaching the consumer. For this reason $^{60}$Co- $\gamma$-rays do not induce radioactivity in foods, and that below 10 MeV energy level, electrons do not produce measurable induced radioactivity even at very high doses. However, electrons with high energies (>20MeV) do produce measurable radioactivity, especially isotopes of Na, P, S & Fe.

### *(ii)Radiological Safety of Operating Personnel*

Being an important consideration, hence has been satisfactorily resolved in countless installations using radiation, including hospitals using x-rays and radioisotope therapy, industrial installations processing isotopes, industrial installations using radiation for cross-linking of polymers and sterilization of medical supplies. The safety system in food industries considers the following factors.

(1) Adequate interlock systems automatically preventing irradiation when personnels enter the radiation chamber.
(2) Adequate shielding to prevent radiation leakage.
(3) Continuous monitoring of radiation at all pertinent points.
(4) Training of personnel in radiation safety procedures.

### *(iii) Danger of Changing Microflora*

This problem is of potential consequence when partial destruction of a microbial population (Pasteurization) rather than sterilization is carried out. This applies to thermal pasteurization and chemical sterilization as well as irradiation. The reason being that in non-preserved foods, there is a certain correspondence between growth of smell-or discoloration - producing bacteria and the growth of bacteria capable of producing disease. This natural correspondence may be disturbed by the preservation treatment so that an innocuous appearing food may become dangerous. Obvious fallacies like that non-processed foods have a built in warning system had led to suspect the safety of irradiated food. However tests showed that irradiated foods are perfectly safe and the suspicion was unfounded.

### *(iv) Nutrient Losses*

The effects of ionizing energy on nutrients are of the same magnitude as those of other processes hence dietary measures like menu planning and nutrient labelling do consider these processing losses.

## Chapter 3

# IRRADIATION SOURCES

Irradiation sources for food processing are either radioisotopes or machine sources. Radioisotopes presently considered for commercial processing are $^{60}$Cobalt and $^{137}$Cesium but in India only $^{60}$Co is employed whose details are as under:

$^{60}$Co is produced by irradiation of $^{59}$Co in a nuclear reactor. The usual procedure is to machine cobalt metal into shapes suitable for the ultimate source design encapsulating it in a material that does not become radioactive during neutron irradiation, and then exposing it to the intense radiation in the reactor. A second encapsulation follows the irradiator, and this must be performed by remote control since exposure of man to the now radioactive cobalt results in certain death Sophisticated remote control method for production, inspection, shipment and installation of $^{60}$Co have been developed and are in use through the world.

$^{60}$Co has a half-life of 5.27 years and emits two $\gamma$-rays per disintegration with energies of 1.17 & 1.33 MeV.

The sources are produced in various shapes having different intensities depending upon their duration of irradiation. There are many variants of the $^{60}$Co irradiation, depending on the desired medical or laboratory research or industrial applications.

Irradiation of a product from two sides with $^{60}$Co $\gamma$-rays has the advantage of providing a fairly uniform dose through out the irradiated food containers *at an energy level which precludes formation of induced radioactivity.*

Board of Radiation and Isotope Technology (BRIT) a constituent unit of the Deptt. of Atomic Energy, established to promote the applications of radiation and isotope technology, has set up a commercial demonstration plant for radiation processing of spices. This plant is located at Vashi, Navi Mumbai – close to the wholesale spices market. It will process, to start with, about 3,000 tonnes of (variety of) spices a year. This will go up to 12,000 tonnes per year in the next couple of years, with argumentation of $^{60}$Co depending on the market demand.

## Technical Features of the above Plant

(i) It is designed for round the clock, fail-safe automatic operation. It is endowed with the flexibility in processing a range of whole and ground spices.

(ii) Tote (Product) box size is 59 cms long, 45 cms wide and 110 cms high. In other words its capacity is $59 \times 110 \times 45 = 0.29$ cu. meter.

(iii) Nearly 30 tonnes of spices can be processed / day at maximum $^{60}$Co loading = one million curies.

Sealing in impermeable packaging materiel is required to maintain quality, prevent infestation and contamination after radiation processing.

Radiation processed spices do not become radioactive just as our hand does not starting emitting X-rays after being X-rayed.

## Machine Sources of Irradiation

Several electron accelerators were employed earlier, however, the most advanced machine food irradiation is the linear electron accelerator. A schematic representation of electron acceleration in the electrostatic accelerator and the linear accelerator is as given on page 25.

One of the disadvantages of electrons is their low penetration as compared to $\gamma$-rays. This can be overcome in part by using high energy electrons, provided the energy is below the level capable of inducing radioactivity. It is understood that electrons with energies below 10 MeV fulfill these conditions. By irradiation from two

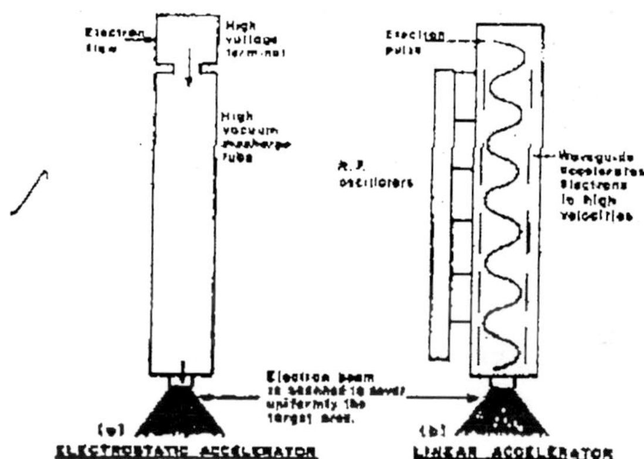


*Schematic representation of electron acceleration in the electrostatic accelerator and the linear accelerator. (a) A high static voltage accelerates electrons down a discharge tube. (b) An electron pulse is accelerated as it travels down the beam tube. The electrons are trapped in a travelling rf wave and accelerated to high speeds.*

sides it is also possible to maximize energy utilization by having a fairly uniform intensity, compared to irradiation from one side only.

## Food Irradiation Physics

In the foregone pages food irradiation chemistry and its effects on the irradiated food have been described. It is worthwhile to be quite familiar with irradiation physics also.

A.K. Jain, *Chemical Weekly*, Dec 5, 2000 pp. 149-159 provides some good details whence the following lines have been drafted.

$^{60}$Co is the source of energy used in a radiation processing facility. $^{60}$Cobalt emits $\gamma$-rays. $^{60}$Co isotope has 5.5 years as its half life and emits 2 $\gamma$-rays of 1.17 and 1.33 MeV. These have high penetration and are easily available. The products, either in packages or in bulk containers enter the irradiation chamber on conveyor system. The source rack with its $^{60}$Co pencils is raised from the storage pool by remote control. The product packages are then moved around the radiation source in such a way that they are exposed to radiation for predefined time. Dosimeters that measure the amount of radiation are placed with the product being irradiated. The treated product is brought out by the conveyor system.

The unit of radiation energy absorbed by product measured by a Unit called gray (Gy) = 1 joule of energy by a kg product.

$\gamma^{60}$Co irradiation has the advantages of reliability, flexibility, effective on all microorganisms without rise of temperature. It is consistent, reproducible, straight forward to validate and to control.

The food irradiation facility generally consists of 4 major parts:-

(1) A source of $\gamma$-radiation ($^{60}$Co).
(2) A shield room to expose and store the source safely–a concrete cell and waterpool.
(3) Mechanism to convey products in and out of the cell and deliver the required dose effectively to the products i.e. the Product Conveyor and Control Mechanism and
(4) Systems to ensure safety and reliability of the operation of the facility- safety devices and inter locks.

## Siting of Irradiators

Geological features that could adversely affect the integrity of radiation shields should be evaluated, taking into account the physiological properties of the materials under lying the irradiator sites or its environs.

In areas with significant potential for seismic disturbance, $\gamma$-irradiators may be equipped with a seismic detector which on being activated by any disturbance will trigger appropriate control mechanism to shield source automatically. In locating the irradiator a minimum screening distance as indicated below needs to be maintained

(a) ammunition, explosive dumps, civil and military airfields–2 kms.
(b) for all other sites – 75 ± 25 meters.

## Codex Alimentarius General Standard

It has been generally accepted and has been adopted that only the following radiation sources are suitable for radiation processing of food: -

(i) Radioisotope sources – $^{60}$Co or $^{137}$ Cs.
(ii) Machine sources:- electron up to 10 MeV and X-rays from electrons up to 15 MeV.

Radioisotope $^{60}$Co production has been detailed earlier. The radioisotope $^{137}$C is obtained from spent fuel elements, but is not readily available in the quantities that would be needed for commercial exploitation.

The quantum energies of the γ-rays from both of these acceptable radio active sources, 0.66 MeV for cesium and 1.13 and 1.33 MeV for $^{60}$Co are well below the threshold for photo nuclear activation of any chemical elements. Consequently even at the highest imaginable doses, no radio activity can be induced in the expased food by these sources.

As indicated above, electrons from machine sources are limited in energy to 10 MeV, and primary electrons for producing X-rays are limited in energy to 5 MeV.

**Chapter 4**

# FOOD IRRADIATORS–THEIR DESIGN AND FABRICATION

## Preamble

Generally food irradiation processing requires the similar type of infrastructure as needed in other physical processes like canning, freezing, drying etc. For these purposes the processing facility must be located at a control point where sufficient quantities of food are produced and transported to the plant for treatment and storage before sending it out to the sale markets. Any treatment of food would add cost to the product. Irradiation technique needs high capital and requires a critical minimum capacity of the irradiation facility for economic operation. But unlike other physical processes, irradiation has low operating cost, especially with regard to the energy required for treating food (For details – Chapter. 12 of this monograph).

Design and fabrication concepts of food irradiation facilities basically focus at providing compact and optimal energy efficient designs. The diversity of irradiators application and consequent design specifications lead to unlimited types of possible irradiators. Consequently conflicting demands crop up, an optimized approach becomes difficult if not impossible.

An irradiator designer usually starts with a given set of parameters, *viz.*

(i) amount of product to be treated,

(ii) the treatment desired in a single or more shifts / day; no of days / week and weeks/year,

(iii) density and dimensions of the products.

From these he calculates the quantum of source activity needed for treatment, followed by the geometric configuration between the source and the target to get an efficient and economical working irradiator. Under PFA Rule 74 – Dose of Irradiation specifies the minimum and maximum dose for each permitted food, hence these greatly influence whether the product can be treated loose or in cartons or pallet quantities.

Two types of irradiator systems are generally relevant to this requirement :–

(i) portable or semi portable units for practical demonstration of the irradiation process as well as for market studies,

(ii) fixed irradiator for techno-economic evaluation and commercialization of the process.

The designer should also take into account, among others the following considerations: -

(a) low capital cost,

(b) plants of low to medium through put,

(c) low marginal cost for the processing,

(d) seasonality of product processed,

(e) technological simplicity and labour involvement,

(f) low operational and maintenance cost,

(g) minimum operator qualifications and training and,

(h) high safety standards.

A major part, over 90% of the capital cost of irradiation facilities is concerned with.

(i) Radiation source

(ii) Biological shield and

(iii) Product handling and source mechanism.

Reduction in capital cost is, therefore, attempted by the appropriate selection and design of these components. The parameters are obviously interdependent variables and an iterative approach is essential to get the optimum value of these factors affecting the overall cost of the irradiator.

The particular irradiation facility to support a given radiation application depends upon

(a) Physical state of the target material.

(b) Target material- in bulk or in package form.

(c) Necessary environmental condition for the target material depending upon its chemical properties.

(d) Product through put.

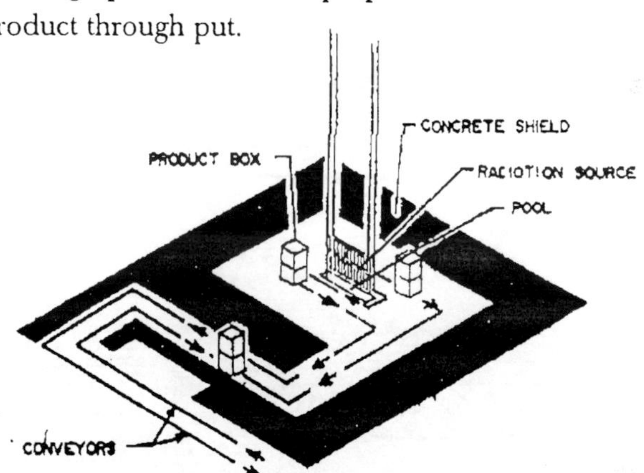


Fig. 4.1 : Schematic of a Gamma Irradiator

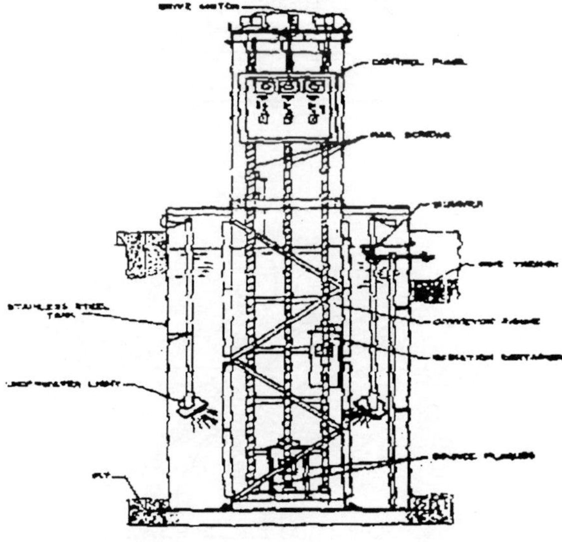

Fig. 4.2 : Research Irradiator

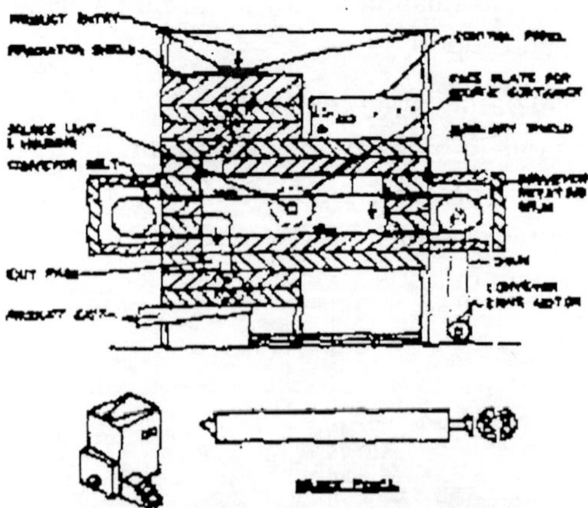

Fig. 4.3 : Transportable Irradiator

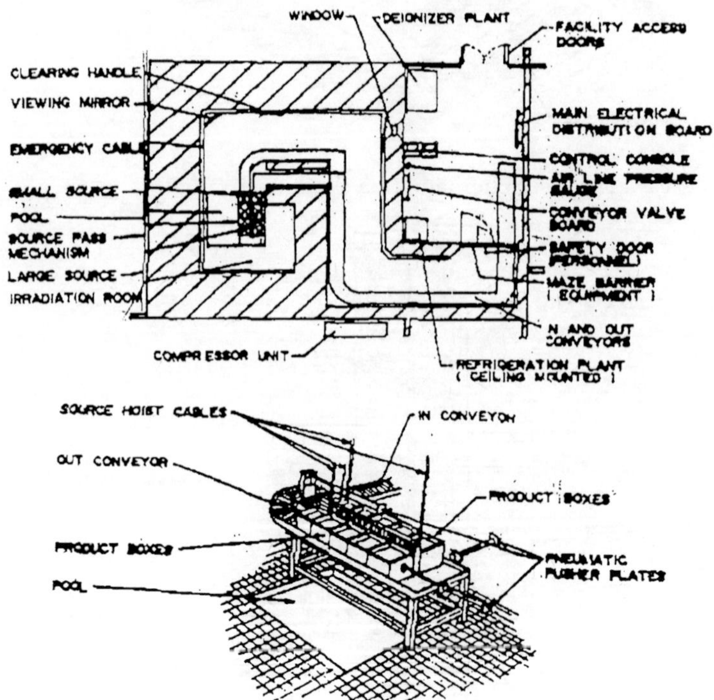


Fig. 4.4 : Semi Pilot Scale Food Package Irradiator

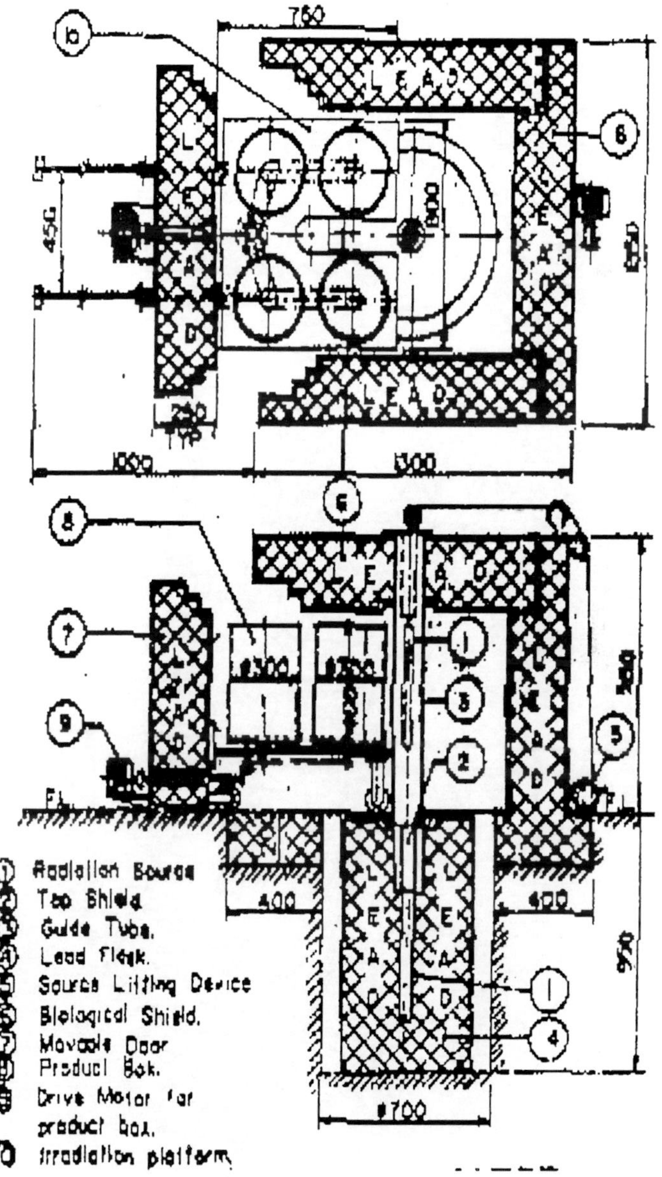


**Fig. 4.5 : Batch Irradiator**

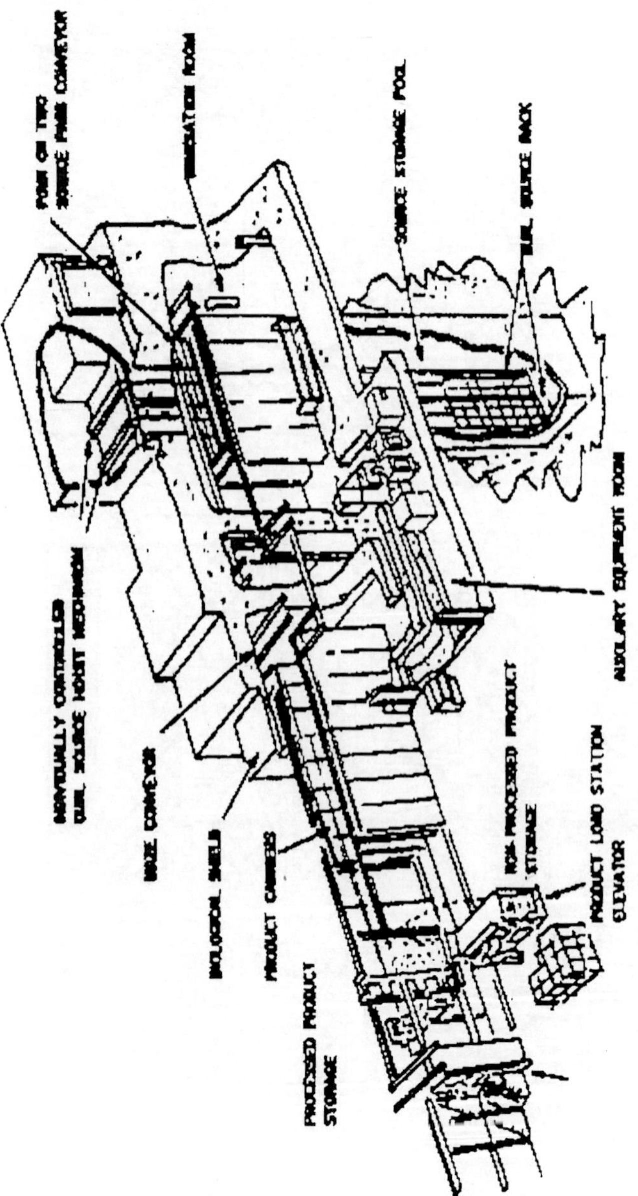


**Fig. 6 : Multi-Purpose Commercial Food Irradiator.**

Chapter 5

# IONIZING RADIATIONS— MAJOR TYPES

## Introduction

Radiation is an energy from travelling through space (radiant energy in a wave pattern and can be either naturally occurring (e.g. from the sun or rocks) or produced by man made objects (e.g. microwaves and television sets). The frequency or wavelength of the energy waves produced by different sources distinguishes the different types and functionality of radiations with high frequency radiation of ultra violet (UV), X-rays and gamma γ-rays posing the most significant risk to human health.

Radiation is called IONIZING radiation when it is at a sufficient high frequency (gamma rays and X-rays) that it results in the production of charged particles (ions) in the material that it comes in contact with. NON-IONIZING radiation, such as that from microwaves, does not produce ions but can create heat under moist conditions and is routinely used for purposes such as cooking and re-heating of foods.

Ionizing radiation can be produced by several sources such as x-rays, electron beams (generated by electron acceleration) or γ-rays. Electron beams are the most cost efficient form of irradiation but they can only penetrate food to a limited depth while X-rays are expensive but are a penetrative form of radiation suitable for bulk operations. However, γ-rays are relatively inexpensive and highly penetrative making them a cost efficient option for any food irradiation. Detailed discussion (chapter 12).

Several radiation types have the characteristic ability to ionize individual atoms or molecules, thereby producing electrons and a +tively charged ion.

$$M \rightarrow M^{+} + e^{-} \quad \text{(i)}$$

Where M is an atom or molecule and the symbol → indicates the action of radiation. Among the important radiations of practical interest in food preservation are electromagnetic waves, including X-rays and $\gamma$-rays; electrons, including $\beta$-rays and cathode rays; and $\alpha$-rays. Of the various ionizing radiations only electro magnetic waves and electrons appear suitable for food processing. Neutrons, deuterons and $\alpha$-particles cause too much damage to the food and also have a significant potential for inducing Radioactivity.

## Electro Magnetic Radiations (EMR)

Light, X-rays, and $\gamma$-rays are all part of the EMR spectrum, which also includes infrared radiation (IR), UV-radiation and radio waves. EMR possess a dual character. According to quantum theory, they may be considered as waves with a characteristic frequency ($\nu$), a wave length ($\lambda$), and a velocity which is independent of ($\nu$). In vacuum, the velocity = $3 \times 10^{10}$ cms/sec. They are also, and with equal justification, considered as packets of energy, or photons. Energy, wave length and frequency are related as :

$$v = C/\lambda, \text{ and} \quad \text{(ii)}$$

$$E = hv, \quad \text{(iii)}$$

where

C is velocity of light in vacuum (cm/sec),

v is frequency ($sec^{-1}$),

$\lambda$ is wavelength (cm)

E is energy of photon (ergs) and

h is Plank's constant ($6.626 \times 10^{-27}$ ergs sec.)

Energy of radiation is often measured in electron volts (eV). One eV is the energy acquired by an electron falling though a potential of IV (1/eV = $1.602 \times 10^{-12}$ ergs; 1/keV = $10^{3}$ eV, MeV = $10^{6}$ eV).

The total effect of EMR upon a material depends upon the energy level of each photon and on the number of photons impinging on the material. Evidently, the lower the ($\lambda$), the higher the energy per photon.

The types of inter actions between matter and various EMR depend on the E level of photons. X-rays and relatively low-energy $\gamma$-rays (EMR with energies of $5 \times 10^4$ eV to $3 \times 10^6$ eV) dissipate their energies in passing through matter by ionization through the so-called "photon electric effect" and the "Compton effect". These effects contribute to total ionization by $\gamma$-rays as a function of $\gamma$-ray energy. At high $\gamma$-ray energies, energy dissipation through "pair production" is possible (the $\gamma$-ray energy is converted into formation of two particles, a negative electron and a positive positron), but this effect is of little importance at $\gamma$-ray energies applicable to food processing.

The photo-electric effect consists of absorption of the photon energy by an atom, with the resulting ejection of an orbital electron. The photon disappears. The energy of the photon is expended partly in overcoming the binding forces holding the electron and is partly converted into kinetic energy of the ejected electron.

$hv = \phi + 1/2\ mv^2$ where, (iv)

$\phi$ is binding energy of the electron (ergs),

m is mass of the electron (g) and

v is the velocity of the electron (cm/sec)

In the Compton effect, the photon behaves as a particle undergoing collision. The electron is ejected with a kinetic energy derived from the photon and the photon then continues with a reduced energy

$hv = \phi + 1/2\ mv^2 + hv'$ where (v)

v' is the frequency of the scottered photon.

Despite the existence of the above three distinct modes of interaction, the absorption of X-rays and of $\gamma$-rays by matter can be represented by the following single exponential equation :

$I = I_0 e^{-\mu} X$, where, (vi)

$I$ is the intensity of radiation at distance X within the absorber

$I_0$ is the intensity of the surface of the absorber

$\mu$ is the co-efficient of absorption. This depends on the frequency of radiation and on the nature of the absorber.

The coefficient of absorption depends on the density of the absorber and is, in fact, almost directly proportional to density for most substances. Coefficients of absorption for a few materials are as under.

| *Materials* | *Absorption Coefficient (cm)* | | |
|---|---|---|---|
| | *0.5 MeV* | *1.0 MeV* | *4.0 MeV* |
| Air (STP) | $8.0 \times 10^{-5}$ | $5.0 \times 10^{-5}$ | $4.0 \times 10^{-5}$ |
| Aluminium | 0.23 | 0.16 | 0.082 |
| Iron | 0.63 | 0.44 | 0.27 |
| Lead | 1.7 | 0.77 | 0.48 |
| Water | 0.09 | 0.067 | 0.033 |

The exponential nature of equation (vi) results in an intensity attenuation pattern. The intensity never reaches zero. Each additional equal thickness of absorber reduces the intensity of the same factor.

A useful value is the "half thickness", as the thickness of the absorber which reduces the intensity of radiation by a factor of two. The "half thickness" is equal to $0.693/\mu$.

## High-Energy Electrons

Electrons emitted from a hot cathode can be accelerated to a very high velocity by several devices which in turn produce ionization effects similar to those produced by $\gamma$-rays. The relation between electron energy and velocity is shown on page 39.

In addition to the acceleration in man-made devices, high energy electrons are also produced by disintegration of some radioactive materials like $24_{Na}$ which emits an electron and is converted to $24_{Mg}$. Electrons produced by radioactive substances are called $\beta$-rays in contrast to cathode rays, which are electron beam produced by man-made machines.

| | *Electron Velocity* | | *Electron Energy* |
|---|---|---|---|
| | *cm/sec* $\times 10^{10}$ | *% of speed of light* | *(eV)* |
| 1. | 0.417 | 13.9 | $5 \times 10^3$ |
| 2. | 1.1 | 36.6 | $5 \times 10^4$ |
| 3. | 2.6 | 86.7 | $5 \times 10^5$ |
| 4. | 2.82 | 94.06 | $10^6$ |
| 5. | 2.985 | 99.57 | $5 \times 10^6$ |
| 6. | 2.994 | 99.869 | $10^7$ |

There is a difference between these two radiations :

Cathode rays can be produced with a uniform energy but β-rays are produced with a spectrum of energies.

Electrons, whether from radioactive nuclei or accelerators, lose energy and are slowed down as they interact with matter, as more and more of the electrons in a beam are slowed down to negligible velocity, the number of fast electrons remaining decrease with depth in an absorber. If a mono energetic beam of electrons enter the absorber the number of fast electrons remaining initially falls linearly with depth. However, not all of the electrons at a given depth have the same energy. Consequently the relative ionization (or energy dissipation) at any point does not vary linearly with distance. In the case of β-rays, which have a distribution of energies, the intensity of ionization varies with depth in a manner to that of X-rays and γ-rays.

The maximum range of an electron beam is related to its energy by the Feather equation:

$$R_{max} = \frac{0.542\,E - 0.133}{Q} \qquad \text{.......(vii)}$$

where,

$R_{max}$ is maximum range (cm),
E is electron energy (MeV) and
Q is absorber density ($g/cm^3$)

This is, however, valid for energies greater than 0.8 MeV.

The design parameters of four major commercial irradiators of our country are listed in Table on page 40.

**Comparison of four major Gamma irradiators of India**

| *S. No.* | *Particulars* | *SARC Irradiator Delhi* | *ISOMED Irradiator Trombay, Mumbai* | *SPICE Irradiator Vashi, Mumbai* | *RASHMI KMIO-Banglore* |
|---|---|---|---|---|---|
| 1. | Radiation source | Co-60 | Co-60 | Co-60 | Co-60 |
| 2. | Loading capacity | 800 KCi | 1000 KCi | 1000 KCi | 300 KCi |
| 3. | Existing Capacity | 447 KCi | 550 KCi | 205KCi | 200KCi |
| 4. | Radiation dose | 10-25 kGy | 25 kGy | 10 kGy | 10-25 kGy |
| 5. | Maximum output | 16,000 tones / year (based on 25 kGy) | 20,000 tones / year (based on 25 kGy) | 12,000 tones / year (based on 25 kGy) | 6000 tones / year (based on 25 kGy) |
| 6. | Source storage | Water pool type | Dry pit type | Water pool type | Water pool type |
| 7. | Dimension & weight of product box | 59(1) × 43 (w) × 34(h) cm. 14.5 kg | 59(1) × 34 (w) × 43(h) cm. 14.5 kg | Tote (product box size) 60(1) × 45(w) × 110(h) cm, 50 kg | Tote boxes 50 kg |
| 8. | Over dose ratio | 1 : 1.2 | 1 : 1.2 | 1 : 1.4 | 1 : 1.4 |
| 9. | No. of boxes in each carrier | 5 | 5 | 2 | 2 |
| 10. | No. of passes | 2 + 2 | 4 + 4 | 2 + 2 | 2 + 2 |
| 11. | Conveyor system | Continuous / Automatic | Continuous / Automatic | Continuous / Automatic | Continuous / Automatic |

**Table : *Contd.***

| *S. No.* | *Particulars* | *SARC Irradiator Delhi* | *ISOMED Irradiator Trombay, Mumbai* | *SPICE Irradiator Vashi, Mumbai* | *RASHMI KMIO-Banglore* |
|---|---|---|---|---|---|
| 12. | Conveyor speed | 1 : 12 (Cell area & labyrinth) | 1 : 4 (Cell area & labyrinth) | 1 : 10 (Cell area & & labyrinth) | 1 : 6 (Cell area & labyrinth) |
| 13. | Radiation safety | Multiple interlocks | Multiple interlocks | Multiple interlocks | Multiple interlocks |
| 14. | Access in to cell | Labyrinth (2 turns) | Labyrinth (3 turns) | Labyrinth (3 turns) | Labyrinth (3 turns) |
| 15. | Personnel access door | Steel | Concrete | Steel | Steel |
| 16. | Control panel | PLC based computerized | PLC based | PLC based computerized | PLC based computerized |
| 17. | No. of carriers | 25 | 31 | 63 | – |
| 18. | Cell area & Volume | 6.8 × 4.2 × 3.5 meters (100 $m^3$) | 7.2 × 6.2 × 3.7 meters (170 $m^3$) | 7.5 × 4.6 × 4 meters (138 $m^3$) | 6.8 × 4.2 × 3.5 meters (100 $m^3$) |
| 19. | System Power | Electric & hydraulic<br>15 KVA, 415 V | Electric & hydraulic<br>– | Electric & hydraulic<br>– | Electric & hydraulic<br>– |
| 20. | Biological Shield | Concrete | Concrete | Concrete | Concrete |
| | Walls | 1.7 m | 1.8 m | 1.8 m | 1.8 m |
| | Roof | 1.5 m | 1.6 m | 1.6 m | 1.6 m |
| 21. | Source geometry | Product overlap | Product overlap | Product overlap | Product overlap |
| 22 | Plant fabricators | BRIT/Symec Mumbai | Marsh Irradiators U.K. | Mather & Platt Pune | BRIT/Symec Mumbai |

## Chapter 6

# DOSIMETRY

It is evident from the effect of radiations with matter that the amount of energy absorbed by matter and the amount of ionization produced depends on several variables, including number of particles passing through an absorber, energy of the particles and the nature of the absorber. Measurement of the radiation dose received by matter is therefore a complex matter.

The International Commission on Radiological Units proposed the term "Rad" – a quantity of radiation which results in absorption of 100 ergs/g at the point of interest.

Curie (cu) is a different kind of unit basic to radiation measurement. It is defined as $3.7 \times 10^{10}$ disintegrations/sec. Since it does not consider the number and the energy of the particles emitted in each disintegration, obviously, two radioactive materials with the same activity as measured in curies may produce very different fluxes.

Several methods are available for detecting and measuring radiation. When the primary purpose is the detection and visualization of tracks of individual radioactive particles, the cloud chamber technique or darkening of photographic films are deemed useful approaches. The cloud chamber has little relevance to the measurement of intensities useful in radiation processing of foods and photographic film is useful as a personnel monitoring device. But photographic film and its response to radiation is, however, extremely useful in autoradiography and other techniques utilizing radioactive tracers in foods.

An important series of measuring devices is based on the ionization chamber principle. We know and by definition also, ionizing radiation produces ions of both signs. By filling a chamber with a gas and by establishing in the chamber an electric field between two electrodes, these ions may be collected at the electrodes. Radiation can then be measured, since it results in a pulse of charged particles at the electrodes.

For measurement of radiations absorbed by food during radiation processing the most useful technique involves use of chemical dosimeters. These in turn must be calibrated against same primary standard. Such a standard is often a calorimetric procedure. Measurement of heat is a much less sensitive technique than the collection and amplification of ions but it has the advantage of directly measuring the energy absorbed in the food or in water, which is after all usually the major food component.

The amount of heat produced by $10^6$ rad absorbed by 1 g of water has been calculated to result in a temperature increase of 2.68°C. The chemical dosimetric systems are available in liquid or solid form. Among the important types are oxidation-reduction systems, such as certain metal organic ions, and various transparent solids, including plastics and glass which darken on exposure to radiations. Some of the solid dosimeters develop color upon irradiation but this color fades later.

**Characteristics of Some Dosimeters**

| *S.* | *Dosimeter* | *Reaction* | *Applicable dose range* | *Precision (%)* |
|---|---|---|---|---|
| 1. | $FeSO_4$ solution | Oxidation | $10^5 - 10^7$ | ± 5 |
| 2. | Methyleneblue solution | Reduction | $10^4 - 10^7$ | ± 10 |
| 3. | Cobalt glass | Darkening | $10^4 - 2 \times 10^6$ | ± 2 |
| 4. | PVC film | Darkening | $10^6 - 10^7$ | ± 15 |
| 5. | Cellophane | Darkening | $10^6 - 10^7$ | ± 15 |
| 6. | Radioluminescence of bibenzyl | Luminescence | $10^5 - 10^7$ | ± 10 |

## Dosimetry and Process Control

Control of the food irradiation process in all types of irradiation facilities involves the use of accepted methods of measuring the absorbed radiation dose and the distribution of that dose in the product package, and of monitoring of the physical parameters of the process.

Dosimetry is the keystone of the proper irradiation processing. Careful dosimetry is required to ensure that a technologically useful dose has been applied, while maintaining the best possible dose uniformity ratio.

## Classification – Dose/Benefits

1. *Low dose applications (< 1 kGy)*
   (a) Inhibition of sprouting in potato and onion.
   (b) Insect disinfection in stored grains, pulses and products.
   (c) Destruction of parasites in meat & meat products.
   (d) Insects disinfection in fruits and dry fruits.
2. *Medium dose application (1– 10 kGy)*
   (i) Elimination of spoilage microbes in fresh fruits, meat and poultry.
   (ii) Elimination of food pathogens in meat and poultry.
   (iii) Hygienization of spices and herbs.
3. *High Dose Applications (above 10 KGy)*
   (a) Sterilization of food for special requirement.
   (b) Shelf-stable foods without refrigeration.

## Dosimetry

Some of the above processes do not need exact radiation dose, but in others a uniform and accurate radiation dose is essential for getting the desired properties in the irradiated food product. Dosimetry, thus, plays an important role. As most of the food irradiation is carried out in house or at a contract service facility, and the customer will often require, as a quality control measure as well as a statutory compulsion, that the absorbed radiation dose in the material is measured accurately using an appropriate dosimeter and documented.

Absorbed dose in a material is defined as quotient of "*de*" by "*dm*", where "*de*" is the mean energy imparted by ionizing radiation to mass of matter dm

$$D = de/dm.$$

The special name of the unit for absorbed dose is "gray" (Gy)

$$1\text{Gy} = 1\text{ J. kg}$$

## Dosimeter and Dosimetry

Dosimeter is a device or system having a reproducible and a measurable response to radiation which can be used to measure the absorbed dose in a given material. Dosimeters with a known level of accuracy and precision should be used for the validation and routine control of any radiation process followed by proper dosimetric measurement procedures, with appropriate statistical controls and documentation.

Dosimetry is the process of measurement of absorbed dose.

**Preamble :** Radiation dosimetry is a well-established technology that can be used over a wide range of doses and in many anticipated applications. Available ASTM standards, national regulations and certification laboratories bear witness to the applicability of this standardized measuring technology. It is based on distinct scientific principles described in several text books and monographs that are widely available.

There are four levels of dosimetry :

(i) absolute; (ii) reference; (iii) routine; and (iv) indicator.

### *(i) Absolute*

Absolute dosimetry systems are usually operated by metrological institutions and serve the purpose of certifying the physical quantity "absorbed energy dose" and its unit the gray (Gy) with very high accuracy and precision; such efforts are usually coordinated on an international level. The inconvenience of carrying out the required procedures limits their application in industrial radiation processing.

***(ii) Reference; (iii) Routine***

Reference dosimetry systems are used and are calibrated against some absolute standard and then linked to the routine dosimetry system, used in process control. In this way, dose measurements are traceable to national and international standards. Recently, label dosimeters have become available. Such systems change colour or exhibit changes in other easy-to-recognize features after reaching a certain dose level; they are useful in routine dosimetry.

***(iv) Indicators***

Indicators must not be confused with label dosimeters; what they have in common is that both are attached to the surface of the products. Indicator cannot "indicate" a dose value; their usefulness and utility is in indicating that the products emerging from the irradiation have been treated.

## National Standards

In India, chemical dosimetry group of Radiation Standards Section, Radiation Safety Systems Division (RSSD), of BARC, maintains national standards for radiation processing. As per Atomic Energy Radiation Board, periodic dose inter-comparison of $\gamma$-irradiator facilities that are being used for food irradiation is a mandatory requirement. This responsibility is entrusted to RSSD, BARC. Glutamine (spectrophotometric readout) dosimeter (detailed elsewhere) having traceability to NPL (UK) is used as transfer standard for dose assurance in spice irradiator (GRPS), BRIT.

The system of validation and routine control of a radiation process is known as Plant Commissioning Dosimetry is summed up below:

## Plant Commissioning Dosimetry (PCD)

PCD is carried out to validate the radiation process i.e. dose mapping studies are conducted to characterise the radiation facility with respect to the magnitude, distribution and reproducibility of dose delivery. This exercise becomes mandatory in case of changes in source position, source strength and redistribution of source activity in source frame.

PCD is generally carried one in three runs – x, y, z.

The x run provides the information about the source uniformity with respect to the product load in the radiation processing cell. These dosimetry results facilitate the optimum utilization of gamma flux coming out of the source frame.

The y run determines the absorption of dose in various segments of the product container and also provides information about the maximum and minimum dose positions in the container. These results dictate the conveyor speed so that the plant runs to deliver specified dose within limits.

The z run is conducted for the statistical evaluation of the absorbed dose and to determine ultimate over dose ratio in the product container.

All the three runs are performed with container cartons filled with dummy material to simulate the product density.

## Selection of Dosimetry System

At the time of deciding a dosimetry system for dose measurement, a few worth considering points are:–

(i) Range of dose of interest – Independent of dose rate.
(ii) Stability and reproducibility – Independent of environment.
(iii) Ease of calibration
(iv) Easy to prepare
(v) Simple to use
(vi) Inexpensive
(vii) Traceable to standard
(viii) No batch to batch laboratory variation.

Based on the characteristics of different types of dosimeters, they are classified as primary, reference, routine and transfer dosimeters. Their corresponding characteristics and uses are as under.

| S.No. | Dosimeter | Characteristics | Uses |
|---|---|---|---|
| 1. | Primary | Highest metrological quality | National and International standard |
| 2. | Reference | High metrological quality | Reference standard |
| 3. | Routine | Calibrated vs. Reference standard | Routine dose measurement |
| 4. | Transfer | Transportable for comparison | For inter-comparison of dose. |

**ISO—11137 Recommended reference standard and transfer standard dosimeters are :**

**A–Reference standard dosimeters - their particulars.**

| *S.No.* | *Dosimeter* | *Read out system* | *Dose range (Gy)* |
|---|---|---|---|
| 1. | Calorimeter | Thermometer | $10–10^5$ |
| 2. | Alanine | ESR spectrometer | $1–10^5$ |
| 3. | Ceric - Cerous ($SO_4$) solution | UV-spectro photometer/ Potentiometer | $10^3–10^5$ |
| 4. | Ethanol– {Chloro benzene solution} | Colorimetry titration/high frequency oscillometry | $10^2–10^5$ |
| 5. | $FeSO_4$ solution | UV-spectro photometer | $10–4\times10^2$ |
| 6. | $K_2Cr_2O_7$ solution | UV-spectro photometer | $10^3–5\times10^4$ |

**Note :** Most of the above dosimeters may also serve as transfer standard dosimeters.

**Systems used for Radiation Processing Dosimetry**

| *S.No.* | *Dosimeter* | *Readout System* | *Dose Range (KGy)* | *ASTM Practice No.* |
|---|---|---|---|---|
| 1. | Calorimeter | Heat measurement by thermistor or thermocouple | 0.1 to 100 | E 1631 |
| 2. | Ionization chamber | Electrometer | $10^{-7}$ to 0.01 | —— |
| 3. | Ferrous Sulphate | Spectrophotometer | 0.02 to 0.04 | E 1026 |
| 4. | Potassium/Silver dichromate | Spectrophotometer | 2 to 50 | E 1401 |
| 5. | Ceric-cerous sulphate | Potentiometry, UV or Visible spectrophotometer | 1 to 100 | E1205 |
| 6. | Ethanol chloro-benzene | High frequency conductivity | 0.01 to 2000 | E 1538 |
| 7. | Alanine pellets/ films | EPR spectrometer | 0.01-100 | E 1607 |
| 8. | Glutamine | Spectrophotometer | 0.1-50 | — |
| 9. | Clear Perspex HX | Vis. spectro-photometer | 0.1-100 | E 1276 |
| 10. | Dyed PMMA films | Spectrophotometer | 0.1-100 | — |

| *S.No.* | *Dosimeter* | *Readout System* | *Dose Range (KGy)* | *ASTM Practice No.* |
|---|---|---|---|---|
| 11. | Gamma chrome YR | Vis. spectro-photometer | 1-100 | — |
| 12. | Cellulose triacetate films | UV or Vis. spectrophotometer | 10-400 | E 1650 |
| 13. | Radio chromic film | Vis. spectro-photometer or densitometer | 0.1-100 | E 1275 |
| 14. | Gaf chromic films | Vis. spectro-photometer | 0.001-50 | — |
| 15. | Sunna films | Optically stimulated luminescence | 1-100 | |
| 16. | Flat polythene sachet With glutamine powder | Vis. spectro-photometer | 0.1-50 | |

**Classification of Dosimeters based on their instrument accuracy**

PRIMARY STANDARDS

Colorimeter, Ionizing Chambers

↓

(D carbon) ± 1% (2σ)

|

D water ± 2% (2σ)

↓

REFERENCE STANDARDS ± 3% (2σ)

Fricke, Ceric-cerous, $Cr_2O_7$,

Alanine, Colorimeters

TRANSFER STANDARDS ↓

Alanine, $Cr_2O_7$

ROUTINE DOSIMETERS ↓ ± 5% (2σ)

Radio chromic films,

Cellulosetriacetate,

clean and Dyed Perspex.

**Source :** RM Bhat, BARC, Mumbai-85.

## B — Routine Dosimeters – Their particulars

| *S.No.* | *Dosimeter* | *Read Out System* | *Dose Range (Gy)* |
|---|---|---|---|
| 1. | Dyed polymethyl metha crylate | VIS spectrophotometer | $10^3$–$5\times10^4$ |
| 2. | Clear polymethyl metha crylate | UV spectrophotometer | $10^3$–$10^5$ |
| 3. | Cellulose triacetate | UV spectrophotometer | $10^3$–$4\times10^5$ |
| 4. | Ceric-cerous ($SO_4$) solution | UV spectrophotometer/ potentiometer | $10^3$–$10^5$ |
| 5. | Radio chromic dye film, solution, optical waveguide | VIS spectrophotometer/ Optical densitometer | 1–$10^5$ |
| 6. | $Fe^{++}$ – $Cu^{++}$ solution | UV spectrophotometer | $10^3$–$3\times10^4$ |

All these dosimeter systems have their own merits and demerits. One has to select a system of dosimeter that suits to his requirements. But any dosimeter system adopted for routine measurements needs to be standardized, calibrated and traceable to national/international laboratories offering such type of services.

## Calibration of Dosimeter

Generally the calibration of routine dosimeters is done in a fixed geometry gamma chamber of known source strength and dose rate at various locations inside the gamma chamber under electron equilibrium condition. When the gamma radiation source is $^{60}$Co then an approximate electron equilibrium can be achieved in a small irradiator volume with about 0.5 g/cm$^2$ of material. Fricke dosimeter is normally used as a reference standard dosimeter for dose rate determination of γ-chamber.

## Traceability of Dosimeter

Public health authorities responsible for enforcement of regulations of radiation processed foods, manufacturers and users of such products – may these be importers as well as exporters are deeply concerned about evidence regarding the proper treatment of such products based on commissioning and routine dosimetry measurements. Consequently International Atomic Energy Agency (IAEA) initiated a unique high dose inter-comparison programme–

"International Dose Assurance Service" (IDAS) with the following aims.

1. To provide a "Dose Assurance" service to promote dosimetric accuracy in products processed in irradiation facilities of IAEA member states.
2. To provide regulatory health authorities concerned with trade of irradiated products with the confidence that such products have been irradiated to the specified absorbed dose.

Alanine (dispersed in polystyrene matrix) dosimeter is being used for this purpose as transfer dosimeter. Free radicals produced are measured with the help of ESR spectrometer. This system was chosen for its consistent response, stability and wide dose range (10 Gy to 100 kGy). More than 10% deviation between the IAEA system and the facility operator's routine dosimeter needs investigation.

## Product Dose Mapping

This study is accomplished by placing the dosimeters throughout selected product load to identify the locations of minimum and maximum dose positions and examine dose uniformity. This information forms the basis while selecting the monitoring locations for routine processing.

## Cycle Time Setting

It is always better hence advised to set the radiation processing parameters, or cycle time, taking into consideration the statistical variations; so that a target dose greater than the required minimum is delivered throughout the product load.

## Calibration

Calibration of dosimeters, dose measuring instruments and cycle timer regularly is not only an essential but an important part of quality assurance system hence needs to be observed to ensure good radiation practices.

## Documentation

A well-documented procedure must be followed for (i) plant commissioning dosimetry (ii) product dose mapping (iii) routine product dosimetry (iv) calibration of instruments. This helps any facility offering radiation processing services to build confidence in end users of radiation processed products.

## Summing up

Selection of a proper dosimetry system, with known uncertainties, helps in the selection of proper plant parameters which influence the dose results in homogeneous dose delivery to all the products under processing. Absorbed dose can be corrected for environmental influences. This helps to generate the confidence in the plant operators as well as the users of the contract service facility.

**Source:** G. Sharma, *"National seminar on Applications of Radio isotopes and Radiation Technology in Food Processing and Health Care Products,"* 1999, 22nd Sept., pp 30-33. Shriram Institute for Industrial Research, 19 University Road, Delhi - 110007.

## Results – A function of Sample Preparation

There are several factors capable of interference in getting accurate results. Some of these are :

## A – Chemi Luminescence

### *(a) Introduction*

Workers who work with liquid scintillation instruments have been confronted by a problem that crop up with certain types of scintillation mixtures in which light-producing events take place *not as a result radioactivity of the sample* under test. These light-producing events may happen by one of several types of reactions like photo luminescence, chemi luminescence or bio luminescence.

All these light sources produce *only one photon per event* hence called *single photon events*. This phenomenon gains significance due to "Coincident detection" method used in LS counting. The sample being counted is monitored simultaneously by two photo multiplier tube, and only those events observed by both tubes are counted. The two tubes must each observe an event within some very brief *resolving time* for the event to be considered coincident, and includent in the count. Typical resolving time are of the order of

$20 \times 10^{-9}$ sec. Since one *singles* event releases only one photon, it cannot be observed by both tubes simultaneously hence is not counted.

However, with a sample in which a large number of *singles* events occur, the probability increases that the two tubes will each observe two different *singles* events at approximately the same instant, thus producing a count.

This type of *random coincidence event* can be significant enough to produce erroneous cpm. Lum-Ex correction provides a means of determining when the results of counting are being distorted by random events.

### (b) Sources of Single Photon Events

Some very common sample preparations can cause so many single photon events that hundreds of thousands of these photons are coincident and look like real cpm.

Conditions of some of these samples are :

1. Samples with an alkaline pH.
2. Samples with per oxides (either organic per oxides or hydrogen per oxide used for dissolving polyacrylamide gels or bleaching hemoglobin). Peroxides with alkali solutions cause particularly severe chemiluminescence.
3. Use of tissue solubilizers, especially with emulsifier cocktails. Tissue solubilizers are designed for use with non-aqueous cocktails.
4. Samples, cocktails or vials exposed to sunlight or UV light from sterilization lamps or UV lamps use detecting fluorescent molecules.
5. Plants extracts containing chlorophyll.

### *(c) Reducing Single Photon Events*

Many single photon events can be eliminated from samples, by some very simple techniques. Chemi luminescence from peroxides and alkaline pH can always almost be eliminated by the addition of glacial acetic acid. Usually 100 ul in 10 ml of cocktail is sufficient.

Sometimes, the above treatment works for chemiluminescence from tissue solubilizers. Failing this, the sample should be set aside and monitored every hour until the chemiluminescence is sufficiently low as shown below:

**Reducing Chemiluminescence and Statics.**

| *Time after treatment (minutes)* | *Chemiluminescence** | | *Statics** | |
|---|---|---|---|---|
| | *KOH + $H_2O_2$* | *KOH + $H_2O_2$+ Acetic Acid* | *Untreated* | *Wiped with antistatic cloth* |
| 0 | 8,330 | 18 | 3,717 | 6 |
| 1 | 3,487 | 11 | 268 | 4 |
| 5 | 1,213 | 6 | 193 | 5 |
| 10 | 639 | 4 | 207 | 6 |
| 20 | 195 | 5 | 148 | 6 |
| 30 | 108 | 6 | 122 | 6 |
| 60 | 48 | 6 | 68 | 5 |

* Counts in 0.1 minutes.

The above table shows the counts of a sample with chemiluminescence. This sample has no Radio Activity. The addition of acetic acid has completely eliminated the random coincidence counts after 5 minutes.

While there is no corresponding method of reducing photoluminescence, the effects fortunately decay rather quickly. Generally, exposure of samples to sunlight or UV light is avoided. If photoluminescence is suspected, the vials are made to sit in the counter for 45 ± 15 minutes before counting.

If the source of single photon events cannot be eliminated then Lum-Ex correction (if installed) can subtract all contributions from single events from each isotope window.

The aforementioned statics are a common source of erratic high counts, resulting from a static charge on the surface of the sample vial discharging in the counting chamber. These static charges accumulate on sample vials handled during sample preparation. The instrument equipped with an electronic ion source will mostly neutralize static charges.

Certain climatic conditions, however, may cause such large static buildups that an appreciable charge remains on the vial. The use of plastic vials, specially when handled while wearing latex gloves, is an other source of static buildup.

The presence of statics can be checked by counting one of our own run samples (not an unquenched standard) repeatedly over a period of time. If the count rate is erratic then there is likelihood of the presence of statics.

The following steps are helpful in solving the static problem.

(i) Avoid latex gloves and use instead polyethylene gloves.

(ii) The gloves should be treated with anti-static spray and/or wiped with an antistatic cloth. This should be a softener cloth of the type commonly used with a household clothes dryer. It is particularly useful since it can prevent statics for many days.

(iii) Each vial must be wiped completely with the anti-static cloth. Statics is very localized on a vial therefore the entire vial must be wiped including sides, top and bottom.

**B – Two Phase Samples**

These are those that get separated into two distinct phases before or during counting hence may yield variable and highly inaccurate data from the counting process. Besides, accurate quench monitoring is not possible. For examples, a sample may be thoroughly emulsified when placed in the instrument, but may separate while waiting to be counted. In the hazy samples the problem is more of phase distinguishing. In vials, even a distinct separation would not be visible.

Nevertheless, this phase problem can often be eliminated by using less sample, more cocktail, or switching to a cocktail with better sample holding characteristics. Dilution with water or neutralizing the pH does help.

To ascertain whether the sample is separating into two phases, a simple test helps. A glass vial of the cocktail and the test sample is prepared. The sample is observed over the time and temperature range it is exposed to during counting. If the sample stays as single phase, there is no problem. If it becomes two phase, some of the above outlined procedure should be tried on fresh preparations. The 2 phase Monitor (if installed) flags any sample detected as two phase.

## Counting Filters And Precipitates

Any sample that is not intimately mixed with the cocktail will count with a low efficiency and with unpredictable and varying results. Quench monitors are not accurate, DPM are difficult to calculate, and even comparisons of CPM between samples may give misleading results.

Counting precipitates on filters or as pellets from centrifugation is a very common procedure. It can be used effectively if certain precautions are taken. There are two main problems in getting accurate and reproducible results from filters or precipitates. First, the amount of physical precipitate will affect the CPM recorded. More sample material will absorb more of the radioactive decay events. This beta absorption results in decreased CPM with increasing amount of precipitate. The second problem results from the precipitate dissolving in the cocktail over a period of time. This results in variability between repeat counts and between different samples.

There are two ways to avoid these problems. The less desirable method is to have samples with approximately the same amount of precipitate and to select a cocktail in which the precipitate will not dissolve. While DPM cannot be calculated for these samples, at least the CPM between samples can be compared.

The other, and better, method is to dissolve the sample completely before counting. In the case of proteins or nucleic acid precipitates, this is a very simple procedure. Add enough 0.05 to 0.1 M KOH (100 uL is a good starting volume) to wet the filter or dissolve the pellet. After mixing, add a cocktail such as Beckman Ready-Solv HP. This will emulsify the KOH along with the sample. It is not necessary for the filter to dissolve. This type of preparation is not affected by the amount of precipitate, and DPM calculations are easily done in the usual manner.

## Distinguishing Sample And Instrument Variability

### *Introduction*

The sample preparation problems discussed in this chapter can lead to counts that are too high, too low, that increase with time, decrease with time, or go up and down in a random pattern. This is often mistaken for instrument malfunction. However, there are instrument problems that can cause similar symptoms. This section outlines how sample problems and instrument problems can be distinguished. These tests are not 100% conclusive but they do form good grounds for you to analyze the most common problems encountered and to communicate valuable information to the service representative.

### *Procedures*

First, calibrate the instrument as outlined in Section 2.7. If a note is printed out that calibration is not successful, be sure an unquenched $^{14}C$ standard was used for the calibration. If upon a repeat attempt the calibration is still unsuccessful call an Authorized Beckman Service Representative. If the calibration is successful, proceed.

Set up a rack with the $^{3}H$, $^{14}C$, and background unquenched standards that were supplied with the instrument. Follow these samples with three or four of your own samples that are giving questionable results.

Set up a User Program as follows:

Count Time : 5 minutes
Isotope 1 : $^{3}H$
Isotope 2 : Wide
Data Calculation : CPM
Count Sample : 3 times
H# (if installed): ON
Lum-Ex : ON
$^{3}H$ and $^{14}C$ precision : 1%

The count time may be increased or the precision decreased if the problem you are evaluating takes place over a longer period of time. Count the samples and review the date as explained in the next section.

## *Analysis of Data*

### *Unquenched Standards*

Look at the sample repeat values for the unquenched standards. Since the $^{3}H$ and $^{14}C$ were counted to a 1% error, the cpm should be +/– 1% of the average 95 out of 100 times. The coefficient of variation should be less than 0.01. The Lum-Ex value should be less than 1%. Check the counting efficiencies, H#'s and backgrounds as described in Sections 7.5.4 - 7.5.6..

### *Samples*

The sample repeat cpm's should be stable and within the counting statistics as discussed for the standards. The Lum-Ex values should be less than 1%. The H# repeats for any one sample should be within a fairly narrow range, depending on the value of the H#. In general, lower H# have less variation. In either case they should not vary by more than +/–10 over the entire quench range of the instrument. If you should ever encounter a real H# problem you will find the variations are in the 100's.

The tests could give different conclusions. These are:

1. The standards and your samples both meet the above criteria. This means the instrument is working. There is still the possibility of an intermittent problem that did not arise over the this test period. Further long term study will be required.
2. The standards check out but your samples vary: This means the instrument is working and yoru sample preparation procedures must be reviewed.
3. The standards don't meet the criteria: Regardless of how your samples perform, if the standards do not meet the criteria outlined within some small limits, then the instrument is not performing properly. In this case, a Authorized Beckman Service Representative should be contacted.

## Recent Developments

Radiation processing of food products requires precise control on the dose delivered and the over dose ratio so that it meets with the national/international standards. Dose distribution varies rapidly where the material is in dose proximity from the source and the density of the material to be irradiated.

**A—** S/Shri Manju Parihar, K. Biju and Amiya Sharma of RPAD (BARC) have described the methodology and the estimated dose for the product of density 0.6 gm/cc moving with a constant speed of 5 cm/minute when irradiated with kCi of $^{60}$Co source.

The above $^{60}$Co source is distributed in 16 pencils arranged in a rectangular frame measuring 47.2 cm × 110.3 cm. The above product is in a package measuring 60 cm(L) × 45 cm(B) × 110 cm(H). The nearest side of the package is at a distance of 17.5 cm from the source frame. The irradiated package shuffles from bottom to top shelf and moves in similar fashion at the above stated speed.

The dose rates are evaluated at nine locations per plane (shown in Fig 3.) and for three planes of the package passing through different location using analytical method and Monte Carlo method.

### *(a) Analytical method*

The general equation[1] used for the line source to evaluate the dose is

$$D = \frac{s_1 k}{4\pi} \int_0^l B \frac{e^{-\mu R_1}}{R^2} dx$$

and $B = (1 + C\mu R_1 e^{D\mu R_1})$

Where D = Dose rate in rads/min

l = length of the source

B = build up factor

$S_i$ = the activity per unit length

C and D are constants for build up factor

$\mu$ = the linear attenuation coefficient of the product.

$R_1$ = the distance traversed by the radiation in the product material

R = the distance between the point of evaluation of the dose rate and the source

k = dose conversion factor.

The above equation calculates exactly dose rate due to single line source which takes into account the attenuation and build up due to the product material. The build up factor is introduced in the equation using Berger's formula. The total dose at a point is calculated from all the line sources considering their relative locations. The above equation is modified for different locations of the package. The integral is solved by Simpson's rule using a computer program.

***(b) Monte Carlo method***

A validated Monte Carlo code MCNP4A[2] is used for the purpose. Entire geometry is modeled as shown in Fig.1 as an input to the MCNP using combinatorial geometry scheme. The source particles are sampled from a rectangular frame of the dimensions under study isotropically. The code takes care of the attenuation and scattering within the product material and the scattering from other product boxes nearby. Point detector tallies are used to score the number and energy fluence spectrum and further converted into dose rate by using appropriate mass energy absorption coefficients. Fractional standard deviation of the results are within 5%.

***Method to incorporate speed***

Various planes are considered along the direction of motion of the package, the dose rates are then plotted as a function of distance with respect to the center, which lie in the line of center of the source, for different planes of the package. The typical graph of the dose rates along the line for different levels for the front plane are presented in fig. 2. The area under the curve divided by speed gives the total dose along the line of motion. The doses are added for the different locations of the package in different planes to get the total dose received for the full cycle.

## Results and Discussions

The dose in kGy along the line as shown in fig. 3 is presented in the table 1. The values calculated using analytical method and Monte Carlo method are in good agreement. The variation is between 2-20%. The values of the Monte Carlo are slightly higher than the analytical calculations for the outer points. This could be because Monte Carlo Simulations accounts for the scattered contributions from the nearby packages. The overdose ratio within the irradiated package is 1.714 and 1.628 using Monte Carlo method and analytical method respectively.

## References

1. Radiation Shielding, J. Kenneth Shultis and Richard E. Faw, Prentice Hall PTR.

2. Briesmeister JF, "MCNP- A general purpose Monte Carlo N- particle transport code, Version 4A, Los Alamos National Laboratory Report, LA- 12625, 1993.

**The dose in kGy along the line in the irradiated package as shown in Fig. 3.**

| Dose along line 1 | | Dose along line 2 | | Dose along line 3 | |
|---|---|---|---|---|---|
| Analytical method | Monte Carlo method | Analytical method | Monte Carlo method | Analytical method | Monte Carlo method |
| 8.98 KGy | 11 KGy | 6.52 Kgy | 7.5 KGy | 8.98 Kgy | 11 Kgy |
| Dose along line 4 | | Dose along line 5 | | Dose along line 6 | |
| Analytical method | Monte Carlo method | Analytical method | Monte Carlo method | Analytical method | Monte Carlo method |
| 10.62 KGy | 12 KGy | 6.84 KGy | 7 KGy | 10.62 KGy | 12 KGy |
| Dose along line 7 | | Dose along line 8 | | Dose along line 9 | |
| Analytical method | Monte Carlo method | Analytical method | Monte Carlo method | Analytical method | Monte Carlo method |
| 8.98 KGy | 11 KGy | 6.52 KGy | 7.5 KGy | 8.98 KGy | 11 KGy |

**B** – S/Shri K. P. Rawat, S. A. Khader, M. Assadullah and K.S.S. Sarma of RTDS (BARC) have described the feasibility of utilizing the 2 MeV/20 KW electron accelerator for process optimisation in low dose applications below 30 kGy for food processing, sterilization and semi conductor irradiation applications.

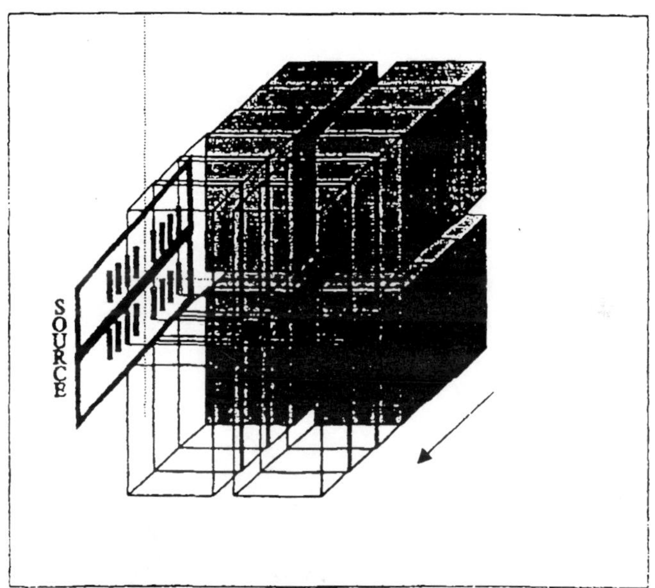


**Fig. 1 : Arrangement of packages during irradiation**

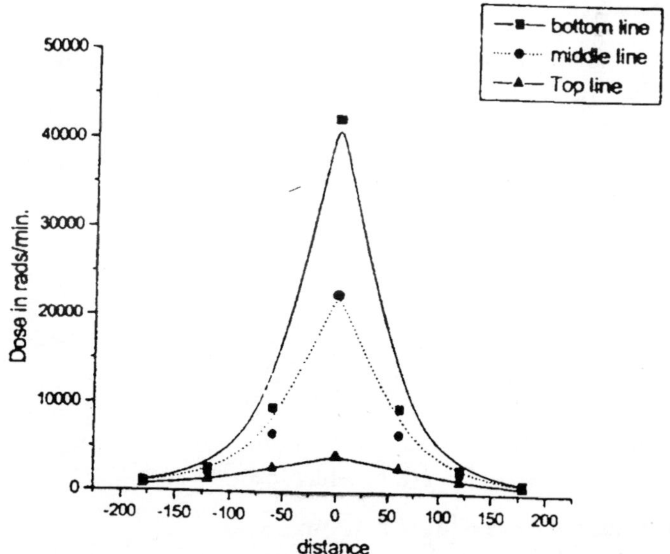


**Fig. 2 : The dose rates along the line of motion of package for the front plane for three levels.**

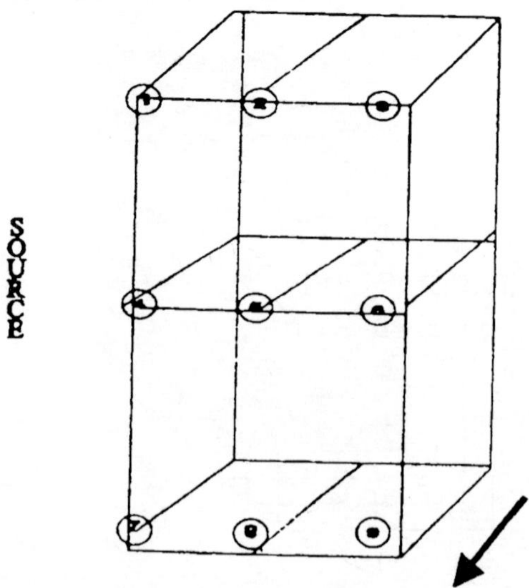


**Fig. 3 : The lines along which the doses are presented in table 1**

## Accelerator

The aforesaid accelerator delivers electron beam pulse in the energy range 1-2 MeV at 50 $H_3$ maximum repetition frequency, with average beam currents upto 10mA. The beam is uniform over a length of 60 cm. Output beam current can be varied by different pulse repetition frequency. Nominal dose rate (kGy/s) have been evaluated from the equation

$D_{av} = I_{av}/A^* (S/p)$, where

$I_{av}$ = average beam curren ($\mu$A)

A = beam area (cm$^2$) and

S/p = mass stopping power (MeV. cm$^2$. gm$^{-1}$) of the medium.

The total dose delivered to the product is a function of average current and conveyor speed on which product is kept.

## Under Beam Irradiation Geometry

A linear mesh conveyor to transport the product in and out of the irradiation zone. The speed can be varied from 0.3 to 6m/

min. The total dose to the product is delivered by sending the same under the irradiation zone in one or more number of times:

Application (Disinfection of wheat flour) from 0.3 to 6m/min. The total dose to the product has been delivered by sending the same under the irradiation zone in one or more number of times.

## APPLICATION

***Disinfestations of wheat flour :*** The recommended dose for disinfestations is 0.25 kGy to 0.5 kGy. To get such low doses, the machine has to be operated under optimum conditions that involve minimum possible pulse current, frequency and higher product conveying speeds, ensuring at the same time that the dose is uniform over the entire product length as it passes through the irradiation zone. The data obtained from the graphite calorimetry at various energies from 1 to 2 MeV at 10 Hz[2] has been used to evaluate the nominal dose for 1 MeV and 2 Hz at conveyor speed 4.5m/min. To ensure the product be uniformly irradiated and to confirm no un-irradiated (unexposed) portions left, fricke dosimetry, a well established dosimetry method has been carried out. Impurity free fricke solution was prepared and placed in ten petri dishes (5 cm dia.) covered with thin PE sheets. The thickness of the solution has been kept to 3 mm which is equal to the useful penetration of 1 MeV electrons in water medium[3]. All the petri dishes with the dosimeter solution were placed on the conveyor one after the other and the same has been exposed to the beam as the conveyor moved with a speed of 4.5 m/min. The change in the optical densities before and after the irradiation were measured. The dose received in each sample has been evaluated. The set up and the data are presented in the figure on page 66. It is evident from the figure that the average dose obtained is 0.30 ± 0.02 kGy which is well within the limits suitable for radiation disinfestations. Wheat flour packets of 1 kg each were made with bulk density is 0.5 g/cc having an optimum thickness 15 mm for double side irradiation. Some live insects (commonly infested) were introduced into the packets and then the packets were hermetically sealed and sent under the beam with set parameters.

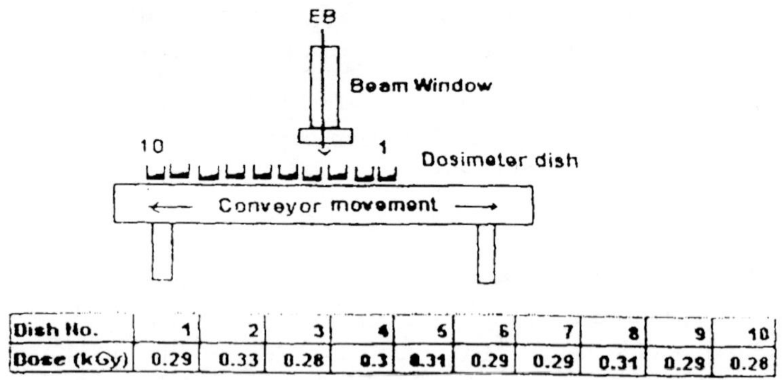


| Dish No. | 1 | 2 | 3 | 4 | 5 | 6 | 7 | 8 | 9 | 10 |
|---|---|---|---|---|---|---|---|---|---|---|
| Dose (kGy) | 0.29 | 0.33 | 0.28 | 0.3 | 0.31 | 0.29 | 0.29 | 0.31 | 0.29 | 0.28 |

Fig.

**Viability of Insects:** No survival fraction of the insects was observed visually over three months.

Thus under these conditions, a typical 1 kg packet with dimensions 35 cm × 35 cm × 1.5 cm, with bulk density 0.54 g/cc can be processed at a througput of approximately 800kg/hour. The throughput can be enhanced and the commercially available 2 kg packets can be directly irradiated if such low currents are available at 2 MeV beam energy, which is possible after some minor modifications in the beam parameters.

***Potato irradiation*** : Feasibility of sprout inhibition in potatoes on electron beam irradiation has been studied. Potatoes were irradiated under ILU-6 EB accelerator under a particular mode where the energy can be as low as 800 keV and the pulse currents are less. This mode of operation cannot be carried out for higher currents. The dose estimated is 0.10±0.01kGy. Since the energy is less, two sided irradiation has been attempted to ensure uniform dose delivery to the potatoes by turning upside down. Some have been exposed only one side to observe the difference between the irradiated and un-irradiated sides. It has been noticed that the inhibition has been achieved in potatoes with double side irradiation. In those irradiated from one side developed inhibition whereas the other side which has not received, showed sprouting. This establishes the fact that sprouting is a surface phenomenon and sprout inhibition can be achieved using low energy electron beam, provided the potatoes are uniformly rotated during exposure.

## Conclusions

The studies established the utilization of ILU-6 EB accelerator at Navi Mumbai for multi product irradiation not only for high dose applications producing value addition to the products, but also for low dose applications like sterilisation of specific products and disinfestations of certain food items.

## References

1. 2 MeV Electron Beam Accelerator at Navi Mumbai – A status Report on the facility characteristics and material processing, K.S.S. Sarma, S.A. Khader, M. Assadullah, S. Sabharwal, L.N. Bandi and D.S. Lavale, Proc. DAE-BRNS INPAC 2003, Indore, Feb. 2003, pp. 292-293.
2. Studies on Beam Characteristics and Dosimetry of pulse electron accelerator, K.S.S. Sarma, A.R. Kalurkar, S.A. Khader, R.S. Deshpande, Proc. thirteenth National Symposium on radiation Physics, Mangala Gangotri, Dec. 1999
3. R.J. Woods and A.K. Pikaev, Applied Radiation Chemistry: Radiation Processing, pp. 20, 1994

# CALIBRATION OF L-GLUTAMINE FREE RADICAL DOSIMETER FOR $^{60}$Co GAMMA RAYS (SPECTROPHOTOMETRIC METHOD)

## Aim

The aim of the experiment is to prepare a calibration graph/relationship between dose and absorbance for L-glutamine free radical dosimeter by spectrophotometric readout method for $^{60}$Co gamma rays. This practice covers the preparation and procedure for using L-glutamine to measure absorbed doses for $^{60}$Co gamma rays. The system consists of L-glutamine powder, an aqueous aerated dilute solution of ferrous ammonium sulphate and xylenol orange (XO) in sulphuric acid (FX) and a high precision spectrophotometer.

## Significance and Use

The L-glutamine (spectrophotometric readout) system provides a reliable means for measurement of absorbed dose in water for

$^{60}Co$ gamma rays, bremsstrahlung photons and high energy electrons. On irradiation of L-glutamine powder, the free radicals produced are trapped in the solid matrix. These stable radicals are estimated by a spectrophotometric read-out technique. 20 mg of accurately weighed powder (irradiated and unirradiated) is dissolved in FX solution. The stable free radicals, on dissolution, oxidise ferrous ions to ferric ions through peroxide reactions. The XO controls the chain oxidation of ferrous ions and forms a complex with ferric ions. The absorbance (A) of the complex is measured spectrophotometrically. The increase in A is not linearly proportional to absorbed dose (D). However a plot of the inverse of A (1/A) is linearly proportional to the inverse of absorbed dose (1/D) upto a dose of 15kGy. The calibration graph between A and D or a mathematical relationship from the inverse graph can be used to calculate unknown doses.

Note. The glutamine system is calibrated against the Fricke reference standard dosimeter[1].

## Requisites

1. Borosilicate or equivalent chemically resistant glassware viz. standard volumetric flasks 500mL and 250mL, stoppered conical flasks 25mL, graduated pipette 10mL and weighing beaker 5mL.
2. Microbalance with 0.01 mgm readability. Calibration of the balance should be checked with a standard weight before weighing reagents.
3. A high precision spectrophotometer capable of measuring A values upto 2 with an uncertainty of no more than 1% in the wavelength region 550nm.
4. Reagents: Ferrous ammonium sulphate $Fe(NH_4)_2(SO_4)_2(6H_2O)$, xylenol orange and sulphuric acid of analytical or reagent grade and single distilled water.
5. Irradiation containers; polystyrene tubes with lids (30mm ht, 1mm wall thickness and 3 mm I.D), 3mm thick build up caps, spacers for centering the irradiation tubes within the gamma chamber, arrangement for determining irradiation temperature.
6. Stainless steel spatula.

## Experimental

### *1. Procedure for cleaning*

a. Glassware: Fill all glassware with a mixture of 1:1 conc. nitric acid and conc. sulphuric acid and keep for 24 hrs. Wash thoroughly with tap water, rinse with distilled water at least 5 times. Dry the conical flasks and the weighing beakers in an over at 100$^0$C for one hr.

b. Polystyrene irradiation tubes. Fill them with a 10% $HNO_3$ solution in water and keep for 24 hrs. Wash thoroughly with tap water, rinse with distilled water at least 5 times. Dry in an oven at 60$^0$C for 1 hr.

## Preparation of solutions

### *a. Stock solution of 2.5 mol.dm$^{-3}$ (5N) $H_2SO_4$ (250mL)*

Add about 125mL of distilled water to a clean 250mL standard volumetric flask. Transfer 30.5mL of $H_2SO_4$ (sp.gr. 1840kgm$^{-3}$) to the flask. Allow the flask to cool. Make up the volume to 250mL with distilled water. Stopper the flask and mix well.

### *b. Preparation of FX solution (500mL)*

Transfer 6.5ml of the stock 2.5mol.dm$^{-3}$ (5N) $H_2SO_4$ to a clean 500mL standard volumetric flask. *Weigh 392.16mgm ferrous ammonium sulphate and 38mgm XO* in two clean dry weighing beakers. Transfer one by one to the 500mL flask. Wash down the remaining contents of the beakers carefully into the flask and *make up the volume to 500mL with distilled water.* Stopper the flask and shake well. Keep it for *30min.* at room temperature. The FX solution has a concentration of $2 \times 10^{-3}$ mol.dm$^{-3}$ ferrous ammonium sulphate, *$10^{-4}$ mol.dm$^{-3}$ XO* in 0.033 mol.dm$^{-3}$ sulphuric acid.

Note: The FX solution will slowly oxidise at room temperature resulting in an increase in absorbance. For accurate results, the FX solution should be used on the same day of its preparation.

## Preparation of glutamine dosimeters

Fill the polypropylene irradiation tubes with L-glutamine with a clean dry spatula. Close the lids, keep in build up caps and number them.

## Irradiation

Place the dosimeters with build up caps in the defined reproducible calibrated position inside the gamma chamber 900. Irradiate a set of atleast 3 dosimeters for each dose value for known timings. (Select the time of irradiation so as to give doses in the range 5 to 15 kGy). Note down the temperature of irradiation. Keep irradiated dosimeters at room temperature.

## Analysis- (Spectrophotometry)

1. Check accuracy of the wavelength scale of the spectrophotometer with a holmium oxide filter. Establish linearity of absorbance scale with a standard potassium dichromate solution (0.055 g/L) in dilute sulphuric acid.
2. Weigh accurately 20 mgm of unirradiated glutamine and transfer to a clean dry conica flask. Close and keep.

Remove the irradiated powder to a clean tracing paper and mix well. Weigh 20 mgm from each sample and transfer to dry conical flasks. A minimum of 3 weighing should be done for each sample. Number the flasks. Finish all weighing.

Rinse a 10 mL graduated pipette with FX solution. Add 10mL FX to the conical flask wit<??> wait for a minimum period of 20 min from the start of addition of FX to unirradiated powder before starting measurements. Use FX solution in the 500mL flask as reference.

3. Set the wavelength control of the spectrophotometer to 549nm.

Clean two matched cuvettes of the double beam spectrophotometer with distilled water. Rinse 2 times with FX and fill them with FX solution. Adjust 0% and 100% transmission. Start measuring absorbance of control (unirradiated glutamine in FX) and samples (irradiated glutamine in FX) against reference solution (FX). Complete all measurements.

Note: Take care to finish measurements within 60 min of adding FX solution to unirradiated/irradiated glutamine powder in conical flasks.

### Dose response relationship

1. Find out average absorbance of control dosimeters ($A_c$) and irradiated samples ($A_{in}$). Calculate net change in absorbance, $\Delta A$. ($\Delta A = A_{in} - A_c$)

1.a. Response correction for irradiation temperature. The response of the dosimeter is independent of irradiation temperature in the temperature range of 23-30°C. Below 23°C, the response decreases by 1.23% per °C decrease in temperature while between 30 and 40°C, the response increases by 0.75% per °C increase in temperature. Above 40°C the response increases by 0.2% per °C. Find out the correction factor required for the irradiation temperature. Apply it to the net change in absorbance ($\Delta A$) for all samples.

2. Calculate the dose given to each glutamine sample. (Multiply absorbed dose rate in water from the Fricke system by time of irradiation).
3. Plot a graph of $\Delta A$ Vs absorbed dose with a suitable fit.

   OR

   Plot a graph of the inverse of DA Vs inverse of absorbed dose with a linear fit. Find out the mathematical relationship between absorbance and dose.

## REFERENCES

1. ASTM Standard E 1026 (94) Standard practice for using the Fricke Reference Standard Dosimetry System.

## DOSE RATE MEASUREMENT IN A GAMMA CHAMBER USING FBX DOSIMETER

### Aim

Calibration of a $^{60}Co$ gamma chamber-900 using the FBX dosimeter.

### Requisites

(1) Spectrophotometer

(2) Cells with 10 mm path length

(3) Standard volumetric flasks 100 ml and 250 ml one each

(4) 10 ml graduated pipette
(5) Spatula
(6) 5 ml beakers-three
(7) 5 ml irradiation tubes made of polypropylene
(8) Irradiation stand
(9) Microbalance
(10) A gamma chamber
(11) ***Chemicals:*** Benzoic acid, ferrous ammonium sulphate, xylenol orange and sulphuric acid [all analytical reagent grade

## Theory

The ferrous sulphate, benzoic acid, xylenol orange (FBX) dosimeter is capable of measuring doses in the range 0.01 to 50 Gy.

The FBX dosimeter consists of 0.2 mol $m^{-3}$ ferrous ammonium sulphate, 5 mol $m^{-3}$ benzoic acid (BA) and 0.2 mol $m^{-3}$ xylenol orange (XO) in 40 mol $m^{-3}$ aqueous aerated sulphuric acid (FBX) system. In the FBX dosimetric system, benzoic acid (BA) increases the $G(Fe^3)$ value. Xylenol orange (XO) controls the BA sensitized chain reaction as well as forms a complex with $Fe^{+3}$. In aerated FBX system each. $H_2OH$ and $H_2O_2$ oxidizes 8.5, 6.6 and 7.6 $Fe^{+2}$ respectively. About 8% OH react with XO and the remaining with benzoic acid. In aerated FBX system part of the H atoms react with $O_2$ and the remaining react with benzoic acid but all H atoms produce $H_2O_2$ through subsequent reactions. The reduction of ferric ions due to H atoms reactions introduces non-linearity of $Fe^{+3}$ versus dose plot in the FBX system.

## Preparation of dosimetric solution :

The FBX solution is prepared using analytical grade chemicals and singly distilled water. To prepare 250 ml of dosimetric solution 152.50 mg of benzoic acid is dissolved in first distilled water in a 250 ml volumetric flask, by warming it on a water bath. 13.9 ml concentrated sulphuric acid is diluted to 100 ml with distilled water 4.0 ml of this diluted acid is added benzoic acid solution. The flask is cooled to room temperature and 38 mg of tetra sodium salt of

XO and 19.6 mg of ferrous ammonium sulphate are added to it. The volume is made up to 250 ml by distilled water. Before adding the ferrous salt to the system, the benzoic acid solution must be brought to room temperature and then acid should be added to the solution.

**Preparation of dosimeters :**

Rinse irradiation tubes at least three times with the dosimetric solution before filling them with solution for irradiation when using a gamma source for irradiation, surround the dosimeters with a 3 mm thick perspex build up material to achieve electron equilibrium condition.

**Irradiation and measurement procedures :**

Dosimeters are irradiated in a gamma chamber at central position. Set the spectrophotometer <??> 100% transmission. The spectrophotometer is set to zero absorbance, measure optical dense irradiated solution.

For Hitachi 100-60 spectrophotometer used in this experiment value of the molar linear absorption coefficient is determined to be 1978 $m^2\ mol^{-1}$.

## OBSERVATIONS

### Optical density (OD) of the blank at 548 nm

| *Sr. No.* | *Irradiation Time (min)* | *OD against solution in the flask* | *Corrected OD* | *Dose (Gy)* |
|---|---|---|---|---|
| 1. | Control | 0.0025 | | |
| 2. | (0.25 min) 15 Sec | 0.247 | 0.2445 | |
| 3. | (0.50 min) 30 Sec | 0.413 | 0.4105 | |
| 4. | (0.75 min) 45 Sec | 0.570 | 0.5675 | |
| 5. | 1 min | 0.727 | 0.7245 | |
| 6. | 1.15 min | 0.812 | 0.8695 | |
| 7. | 1.30 min | 1.020 | 1.0175 | |
| 8. | 1.45 min | 1.156 | 1.1535 | |
| 9. | 2 min | 1.277 | 1.2745 | |

In a graph of time of irradiation versus dose the slope of the straight line drawn gives the rate.

The dose D in Gy is given by the following relationship

$$D = \frac{0.179}{\varepsilon l / A - 0.003} \quad \text{or} \quad D(Gy) = \frac{9.05}{(l/A) - 0.152}$$

where ε is known as the molar linear absorption coefficient. For Hitachi spectrophotometer used in this experiment the value of ε is 1978 $m^2$ $mol^{-1}$ and has been independent of temperature between 25°–45°C. This value has to be determined for of XO used and the spectrophotometer separately.

l = path length of the optical cell in cm.

In a graph of time of irradiation versus dose the slope of the straight line drawn is rate (Gy/min).

## ELECTRON SPIN RESONANCE (ESR) BASED DOSIMETER

**Theory :** Electron spin Resonance (ESR) technique is an important research tool wherein the unpaired electrons are excited under a magnetic field by microwave absorption to a higher energy state. The microwave absorption is studied as the first derivative ESR signal, measured as a function of radical concentration which in turn is proportional to the radiation dose. It can be seen as a ligand nuclear hyperfine structure on the ESR of the nuclear hyperfine structure on the unpaired electron localized on the ligand (a radical ion). Standard reference systems have been developed for ESR dosimeter materials like alanine, quartz and a few other new materials.

**Experimental:** T.S. Sudheer et. al. of BARC, Mumbai-85 have studied the suitability of a 2:1 mixture of lithium oxalate and sodium oxalate for gamma dosimetric investigations and the yield of oxalate radical ion as a function of gamma dose using ESR technique.

On gamma irradiation it has been found to give $CO_2^-$ radical (g = 2.015±0.001) which is stable. The power dependence of the radical signal shows saturation above 2.38 mW from a study of ESR spectra for materials irradiated to different doses. The yield of the radicals is proportional to the radiation dose. The dose response of the material is found to be linear in the gamma dose range 6Gy to 11kGy with good reproducebility and stability.

Obviously this is a possible ESR dosimeter in high range dosimetry for gamma radiation, with good signal stability for a period of 2 months.

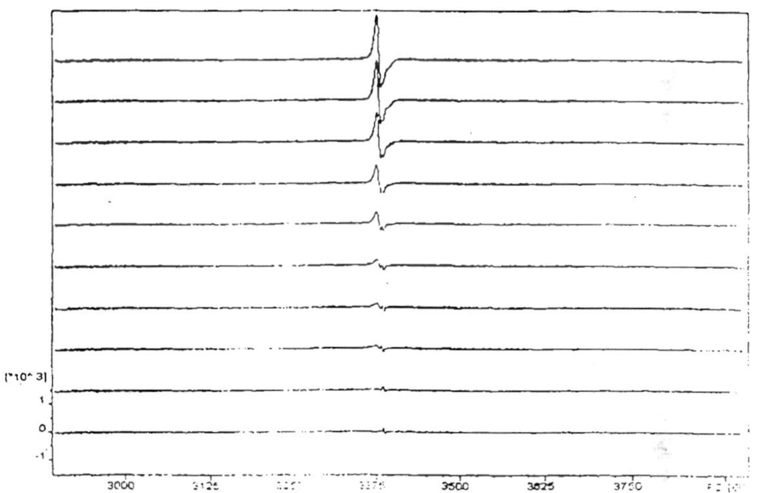


**Fig. 1 : EPR spectra of Lithium & sodium mixed oxalate gamma irradiated to 36 Gy, 73 Gy, 180 Gy, 360 Gy, 1100 Gy, 2200 Gy, 4400 Gy, 6600 Gy, 8800 Gy and 11 kGy samples**

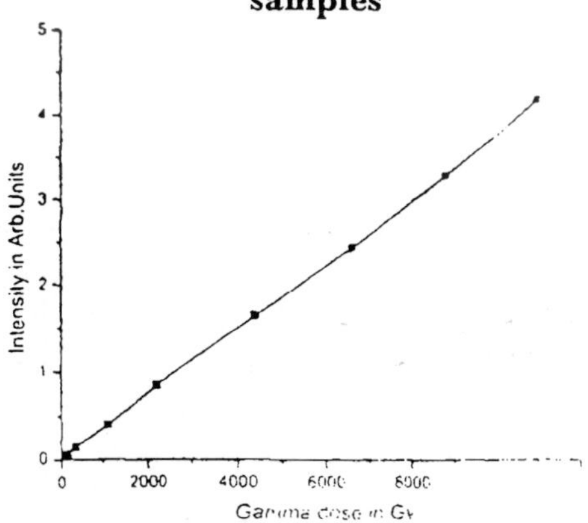


**Fig. 2 : Gamma dose response from 6 Gy to 11000 Gy.**

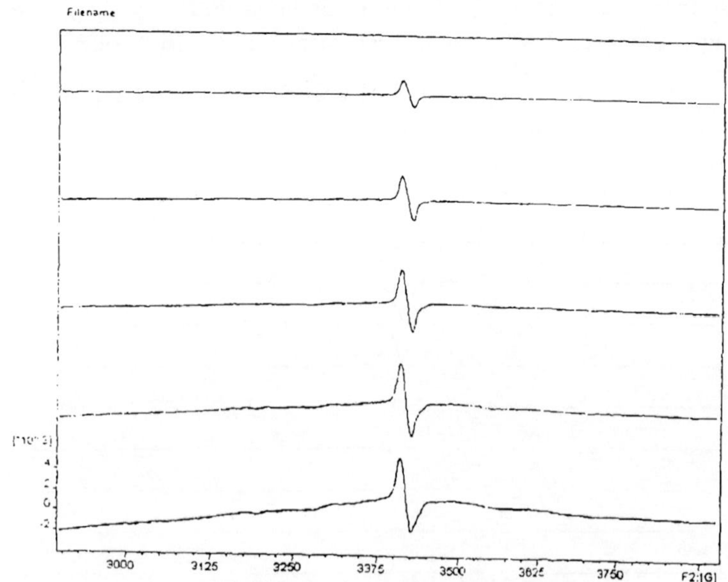


**Fig. 3 : ESR spectrum showing the power dependence 23.7 mW, 7.4 mW, 2.38 mW, 0.75 mW, 236 μW from bottom to top.**

## Chapter 7

# TECHNOLOGY ASPECTS OF RADIATION PROCESSING

Ionizing radiations have applications in several areas of food processing, summarized below :-

(i) Sterilization of foods in hermetically sealed containers is RADAPPERTIZING. It is directed to production of foods stable at room temperature for an indefinite period of time. It may be preferable to canning for those foods which are adversely affected by heating but can tolerate sterilizing dose of radiation.

(ii) Extension of shelf like of various foods to be distributed and stored at refrigerated temperatures like fresh fish, poultry, meats, fresh vegetables and fruit, milk, eggs, and cheese. For many of these the shelf life can he extended by reducing the microbial populations and /or eliminating potentially pathogenic micro-organisms.

(iii) Treatment of water, sewage and food processing wastes. An interesting application of this type is the treatment of cooling water for fish on board ships. By monitoring the microbial count in this water at a low level without heating the water, it should be possible to retard deterioration of fish prior to the time of their delivery for subsequent procesing at the port harbour

(iv) Disinfection of grains or dried fruits and of other commodities subjects to insect attack .

*Inhibition of sprouting*: Onions, potatoes, and carrots are commonly stored extensively at relatively high temperatures for long periods of time. Of course sprouting can be prevented by chemical treatment but the alternative method of ionizing sterilizing requires low dose where chemical effects are negligible unlike the residue from the former chemical treatments.

Other food applications like irradiation of dehydrated vegetables. Irradiation causes scission of material and this scission is beneficial in acceleration rehydration of otherwise poor rehydrating vegetables.

Another interesting application is the possibility of accelerated aging of Scotch whisky. The flavor and appearance of aged whisky are due to oxidative changes which occur slowly under normal conditions but is greatly accelerated by irradiation.

## Chemical Effects of Ionizing Radiations

Ionizing radiations split *i.e.* radiolyze water. Since most foods are of aqueous system, this effect on water is of key importance. The passage of ionizing radiations though water results, first of all in the formation of the following intermediates

(i) Excited water — $(H_2O)^{\cdot}$
(ii) Free radicals — $OH^{\cdot}$ and $H^{\cdot}$
(iii) Ionized water molecules — $(H_2O)^{+}$
(iv) Hydrated election — $e_{aq}^{-}$

These species then react among themselves or with other components of the system. In pure water and in the presence of air they produce in particular the following:-

(a) Hydrogen gas — $H_2$
(b) Hydrogen peroxide — $H_2O_2$
(c) Water — $H_2O$
(d) Hydronium ion — $H_3O+$
(e) Hydroxide ion — $OH^-$

The reactions that these intermediates can undergo with food components are too numerous. Every class of food constituents, including carbohydrates, proteins and other nitrogenous compounds, oils and fats, vitamins, enzymes, and pigments, react with at least some of the intermediates to produce new intermediate compound, many of which are highly reactive themselves. Oxidation reactions, free radical reactions and reduction reactions are particularly significant in this respect. Besides above, radiation has direct and significant effects on organic compounds, specially non-aqueous systems.

In direct action on hydrocarbon chains, several primary events can occur. Hydrocarbon chains are of course present in lipids as well as in many polymers of food packaging materials. The most important event is the abstraction of a hydrogen and the formation of a free radical:

$$-(CH_2)\text{—vvvv}\rightarrow -(\dot{C}H)^- + \dot{H}$$

The hydrocarbon radicals can then undergo a very large number of reactions. Among these, reactions with atmosphere or dissolved oxygen and cross linking are the most important.

## Polymers under Ionizing Radiations

In polymers, as well as in some compound of low mol. wt. scission by the ionizing radiation is also possible. Different groups in polymers have different radiation sensitivities. Aromatic groups in polymers are the most stable, and groups containing halogens (Cl and F) are among the least.

## Proteins under Ionizing Radiation

In proteins, hydrogen abstraction usually results in free radicals on the α-C or on the S-containing amino acid, phenylalanine, and tyrosine are the most susceptible to irradiation.

Deamination, oxidation, polymerization, and decarboxylation have all been implicated in protein changes during irradiation. Reactions of proteins exposed to radiation have similarities to those caused by organic peroxides. In both case free radicals of proteins can either cross link, recombine with hydrogen or result in scission depending on various environmental factors, including the presence of water and oxygen.

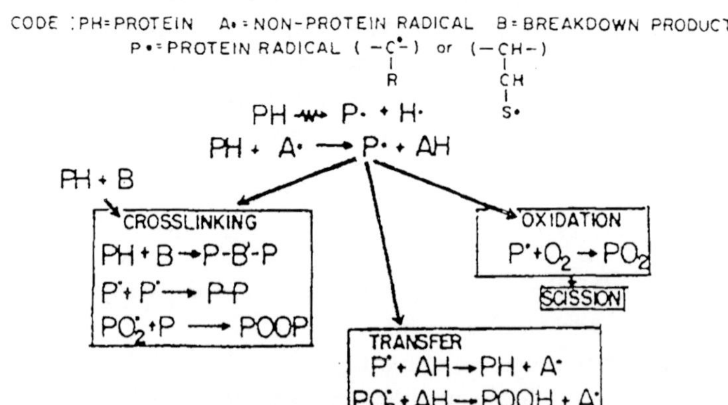


**Radiation-initiated and oxidation-initiated free-radical reactions of proteins.**

## Re-irradiation

Under certain circumstances it might be justified to process a food commodity that has already been irradiated. Rule 76 of PFA Rules 1955 states

(b) "Foods which have been treated by irradiation shall be identified in such a way as to pervent its being subjected to re-irradiation.

(c) Food once irradiated shall not be re-irradiated unless specifically so permitted under these rules.

Similarly under WHO, vide WHO Technical Report Series No. 659 (1981).

The basis of these rules are such situations as:- dry products like grain, irradiated for insect disinfections where re-infestation requires repetition of the treatment, as with fumigation products manufactured from a raw material that has been irradiated for same purpose, much as onions previously irradiated to inhibit or dried onions prepared from irradiated onions, but where the final product needs to be processed by ionizing radiation for some justified purpose; and an irradiated minor ingredient like spices, where the final product containing this ingradient is to be irradiated for a justified purpose.

It was concluded that the additional amounts of radiolysis compounds in the final products would be insignificant and, hence this practice would be acceptable.

**Sources:-** (i) WHO, Technical Repast Series into 659 (1981)

(ii) FAO, CAV/Vol. XV-Ed-1. (1984)

WHO, WHO/EHE/FOS/89.5, available on request for Programmes of Food Safety and Food Aid; WHO, 1211, Geneva 27, Switzerland (1989).

This rationale also applies for products, processed to higher doses by ionizing radiation in order to obtain sterility, that contain irradiated ingredients or raw materials; an example is a composite foods or meal containing vegetables, meats and spices, all previously irradiated for purpose other than sterilization. The dose and incremental amount of radiolysis compounds would be insignificant and hence, the use of previously irradiated ingredients in products to be sterilized by irradiations would not need special processing considerations. Situations other than those that require repeated irradiation are not in compliance with good manufacturing practice and should be considered unacceptable.

It should be noted that fractionated irradiation where the full dose is applied in two or more instatements is NOT considered to be a repeated irradiation; fractionation could occur when the irradiation in interrupted for technical reasons (e.g. failure of transport system):

Concludingly, the overall handling of the product can be sufficiently controlled to ensure that products receive the *required* sterilizing dose, either in one treatment or in a properly fractional sequence of treatments, and that they are not re-irradiated unless technically justified.

Chapter 8

# NUTRITIONAL CONSIDERATIONS

## Commonality and Predictability

Numerous investigations have confirmed the principle of commonality and predictability of radiation effects discussed earlier. Loss of nutrients in cases with radiation dose, but the rate of loss can differ substantially. Factors modifying the effects of radiation, like oxygen, water or temperature will affect different foods to about the same extent. For instances riboflavin is much move stable while thiamine and tocopherols are radiation- sensitive in any food.

Complexities of the radiation chemistry in different foods are not understood in every detail. One finds radiation induced loss of ∝-tocopherol was found to be consistently greater in turkey breast than in beef, pork, lamb or turkey leg, loss of thiamine was found some what more in beef and turkey breast than in lamb, pork and turkey leg. Analysis of sulfhydryl, protein, moisture, fat or water content, pH or reducing capacity by redox titration provided no explanation for those differences in retention. Explanations are hypothetical.

## Macro nutrients

Studies in human volunteers consuming a variety of foods irradiated with a dose of 28 kGy revealed no effects of radiation on metabolizable energy, nitrogen balance or co-efficient of digestibility. Since proteins provide essential amino acids needed for human organism to make its own proteins. A comprehensive toxicological investigation of chicken meat radiation, sterilized with a dose of 50 kGy by electron beam or γ-rays involved the determination of the protein efficiency ratio (PER) of the chicken meat by rat growth assays; no effect of radiation was observed. The amino-acid pattern of the irradiated chicken meat also remained unaffected.

Besides above, protein quality measured as net protein utilization (NPU) is not adversely affected by irradiation.

Foods of plant origin had no effect at a dose level of 28 kGy on the biological value of corn protein or wheat gluten. Irradiation of cereals with high doses was repeatedly found to improve somewhat the nutritional value of cereal proteins as determined in chick growth assays. For example, wheat bran irradiated to 50 kGy had NPU of 40.3% ; 36.0% of non-irradiated bran.

In brief, the evidence from animal feeding studies and from chemical analyses indicates that nutritionally relevant losses of protein quality do not occur in the dose range up to 70 kGy. The absence of any significant trend with dose suggests that even higher doses would not be of concern.

## Vitamins

Life thermal treatments, radiation processing of foods causes some loss of vitamins, As stated earlier, some vitamins are quite insensitive to ionizing radiation, whereas others are rather radiation sensitive.

Diehl (1992) has provided relative radiation sensitivity of vitamins as tabulated below.

RELATIVE RADIATION SENSITIVITY OF VITAMINS

| *Most sensitive* | | | | | | | | *Least sensitive* |
|---|---|---|---|---|---|---|---|---|
| ***Fat Soluble Vitamins*** | | | | | | | | |
| Vit E | → | carotene | → | Vit A | → | Vit D | → | Vit K |
| ***Water soluble vitamins*** | | | | | | | | |
| Vit $B_1$ (Thiamine) | → | Vit C | → | Vit $B_6$ | → | Vit $B_2$ | → | |
| Folate | → | Niacin | → | Vit $B_{12}$ | | | | |

*Source* :– JF Diehl, Food Additives & Contaminants (1992), 9, 409-416.

However this ranking of sensitivities is not always strictly applicable. Several factors influence the radiation resistance of a vitamin. These are:

(i) Composition of the food under consideration.
(ii) The packaging environs.
(iii) The temperature during irradiation.
(iv) The post irradiation storage.

The presence or absence of oxygen in the packaging environs, has a particularly pronounced effect on Vit E. Beef irradiated to 30 kGy under nitrogen, no loss of Vit E was detected but in presence of air, 37% was missing. Similarly chick feed irradiated at 50 kGy posted a loss of only 10% of Vit E under vacuum but by packing in air the loss shot to 51%.

Foods marked for high dose irradiation need processing conditions to mitigate undesirable sensory effects, reduction of vitamin losses is the added advantage. Oxygen must be removed before irradiation, cryogenic temperatures are more favorable to protect vitamins during and after irradiation.

Tables on next page are self explanatory.

Data relating to the effect of subfreezing temperature and of radiation sources demonstrate once again the improved retention of thiamine when irradiation is carried out at lower temperatures and the much better retention of thiamine with electron irradiation than with $\gamma$-ray irradiation due to the much higher dose rate delivered by the electron beam which favours radical-radical reactions over radical–substrate reactions.

Thiamine (Vit $B_1$) is the most radiation – sensitive of the water-soluble vitamins. Irrespective of the assay methods, 65±5% of thiamine in beef was found to be destroyed by a dose of 30 kGy delivered under less than ideal conditions. The aberration were delivered under less than ideal conditions. The observation were:-

(a) samples though sealed in cans but apparently oxygen not excluded before sealing.
(b) Though samples were shipped frozen to the irradiation facility but the temperature during irradiation had not been indicated.

**Vitamin content of frozen, thermally processed, gamma-irradiated and electron-irradiated enzyme-inactivated chicken meat[a]**

| *Vitamin* | *Process* | | | |
|---|---|---|---|---|
| | *frozen control* | *Heat-sterilized* | *Gamma irradiated (59 kGy at – 25°C)* | *Electron-irradiated (59 kGy at – 25°C)* |
| Thiamine hydrochloride (mg/kg) | 2.31 | 1.53[b] | 1.57[b] | 1.98 |
| Riboflavin (mg/kg) | 4.32 | 4.60 | 4.46 | 4.90[c] |
| Pyridoxine (mg/kg) | 7.26 | 7.62 | 5.32 | 6.70 |
| Nicotini acid (niacin) (mg/kg) | 212.9 | 213.9 | 197.9 | 208.2 |
| Pantothenic acid (mg/kg) | 24.0 | 21.8 | 23.5 | 24.9 |
| Biotin (mg/kg) | 0.093 | 0.097 | 0.098 | 0.103 |
| Folic acid (mg/kg) | 0.83 | 1.22 | 1.26 | 1.47[c] |
| Vitamin A (IU/kg) | 2716 | 2340 | 2270 | 2270 |
| Vitamin D (IU/kg) | 375.1 | 342.8 | 354.0 | 466.1 |
| Vitamin K (mg/kg) | 1.29 | 1.01 | 0.81 | 0.85 |
| Vitamin $B_{12}$ (mg/kg) | 0.008 | 0.016[c] | 0.014[c] | 0.009 |

[a] Vitamin concentrations are given on a dry weight basis.
[b] Significantly lower than frozen control.
[c] Significantly higher than frozen control.

To provide microbiologically safe diet items to the immuno suppressed patients, dairy products were packaged under nitrogen and irradiated to a dose level of 40 kGy at –78 °C; yoghurt bars and nonfat dry milk lost about 25% of their thiamine content whereas in ice cream, mozzarella cheese and cheddar cheese thiamine levels were unaffected by irradiation.

The combined effect of irradiation and frying in bacon was more than a simple addition if the bacon was first irradiated and than fried and when bacon was first fried and then irradiated as tabulated on p. 87.

It is explained that frying reduced water content of fried bacon hence thiamine losses were smaller.

**Thaimine in raw, fried, irradiated-fried, and fried-irradiated bacon at three radiation doses and two irradiation temperatures**

| | | Irradiation temperature 2°C | | | | Irradiation temperature −40°C | | | |
|---|---|---|---|---|---|---|---|---|---|
| | | | % loss | | | | % loss | | |
| *Treatment* | *Radiation dose (kGy)* | *Thiamine (mg/100g protein)* | *Due to irradiation* | *Due to combined treatment* | *Due to frying* | *Thiamine (mg/100g protein)* | *Due to irradiation* | *Due to combined treatment* | *Due to frying* |
| None (raw) | 0 | 4.42 | – | – | – | 4.54 | – | – | – |
| | 7.5 | 1.77 | 60[a] | – | – | 3.84 | 15[a] | – | – |
| | 15.0 | 0.95 | 78[a] | – | – | 3.09 | 32[a] | – | – |
| | 30.0 | 0.40 | 91[a] | – | – | 1.70 | 62[a] | – | – |
| Irradiated, then fried | 0 | 2.28 | – | 48[a] | 48[b] | 2.28 | – | 50[a] | 50[b] |
| | 7.5 | 0.76 | – | 83[a] | 57[b] | 1.73 | – | 62[a] | 55[b] |
| | 15.0 | 0.40 | – | 91[a] | 58[b] | 1.34 | – | 70[a] | 57[b] |
| | 30.0 | 0.07 | – | 98[a] | 82[b] | 0.16 | – | 96[a] | 91[b] |
| Fried, then irradiated | 0 | 2.32 | – | 47[a] | 47[b] | 2.18 | – | 52[a] | 52[b] |
| | 7.5 | 2.02 | 13[c] | 54[a] | – | 1.94 | 11[c] | 57[a] | – |
| | 15.0 | 1.78 | 23[c] | 60[a] | – | 1.73 | 21[c] | 62[a] | – |
| | 30.0 | 1.49 | 36[c] | 66[a] | – | 1.48 | 32[c] | 67[a] | – |

[a] Compared to non-irradiated, non-fried samples.

[b] Compared to non-fried samples irradiated to the same dose.

[c] Compared to non-irradiated, fried samples.

Another study confirmed that there is no formation of anti vitamin factors like anti thiamine and anti pyridoxine as a result of irradiation in chicken and beef .

V it$B_{12}$ is quite insensitive to irradiation as tested in hoddock fillets irradiated to 25 kGy, in fishes irradiated to a dose of 30 kGy or in dairy products sterilized with a dose of 40 kGy at –78°C it in a nitrogen atmosphere.

Niacin has low radiation sensitivity. Folic acid does not suffer any loss in irradiation sterilized beef, chick diet., and several food stuffs. Vegetables, the main dietary source of foliates are not irradiated through high doses (> 10 kGy).

***Vit C is a radiation*** – sensitive vitamin at higher doses of irradiation. Fresh fruits and fruit juices, vegetables and potatoes are the most important sources of Vit C in human nutrition. Those source are not subjected to higher does to avoid undesirable change in their sensory qualities hence Vit C is safe there irradiated products.

***Vit A***:- since radiation reduced loss depend the temperature and the atmosphere during irradiation, the results for Vit A in cream cheese are especially instructive. Determination mode 4 weeks after the irradiation of cream cheese to 50 kGy indicated that 5% of the vitamin A was lost when irradiation was undertaken vacuum at ambient temperature 5% with irradiation in air at –80° and 60% with irradiation main at ambient temperature.

Consequently Vit A bearing milk products and also carotene (pro vitamin A) providing vegetables and fruits should not he irradiated of high doses.

***Vit D*** – It is less radiation sensitive than Vit A as show in table (p 42) - p. 86.

***Vit E***. – Main sources of vit E in human nutrition are margarine, butter, vegetable oils and fats. None of these food stuffs present a microbiological problem hence do not need irradiation. However, most high-fat foods suffer undesirable changes in sensory quality when irradiated to high dose.

***Vit K*** : This vitamin is insensitive in vegetables like cabbage, spinach, broccoli and some other vegetables even at 56 kGy levels of irradiation and after storage of 15 months at room temperature.

These fat soluble vitamins are less stable in meat / beef 0°C based foods. That a content of fillets of Jagfish irradiated at 0° to 3 kGy was unaffected; 45% was lost after irradiation with 30 kGy. Vit D is less radiation – sensitive than vit A. Vit E is sensitive to irradiation but is not required in these foods, as stated earlier, due to there immunity from micro biological problem Vit K appears to be less stable in beef. The loss d Vit K in the irradiated chicken meat was about 36%

In short, except for thiamine, the loss d vitamins is following high dose-irradiation of foods is insignificant and not a concern. For thiamine, the impact on dietary intake needs to be considered.

## Polyunsaturated fatty Acids (PUFA)

Following studies inculcate that the irradiation of food in the dose range under consideration of FAO has no or only marginal effects on essential fatty acids

| | | |
|---|---|---|
| (i) | Herring fillets irradiated | 59 kGy dose – no destruction of PUFA |
| (ii) | (a) Whole grains of rice, wheat & rye | 0.1 – 1kGy – no less |
| | (b) Whole grains of rice, wheat & rye | 63kGy – only small less |
| | (c) Soybean (linoleic acid) | 100kGy – no change |
| | (d) peanut kernels | 20 kGy analysed significant after 1 year |
| | (e) chicken meat | (Raltech study) –do- |

***Minerals/trace Elements*** :– These are not affected by irradiation No evidence that the bioavailability of these elements might be adversely affected by irradiation.

***Summing up*** :– (i) Macro nutrients i.e. proteins fats and carbohydrates are not significantly altered nutritionally and their digestibility by irradiation treatment.

## Micro nutrients

(1) Same of the vitamins are susceptible to irradiation to an extent very much dependent upon the composition of the food, processing and storage conditions. Irradiation done cryogenically and in absence of oxygen does not affect the retention of the sensory quality of foods, on the contrary it enhances the retention of quality. Irradiated foods are[3] superior to thermally sterilized foods.

Thiamine in foods is affected by irradiation as a already detailed.

# Chapter 9

# TOXICOLOGICAL CONSIDERATIONS

Earlier when irradiation was suggested as a means of increasing the storage life of agricultural commodities, it was found to increase the susceptibility of some products to certain storage fungi especially *Aspergillus spp*, mainly *A. parasiticus* and *A. flavus* which leads to aflatoxin contamination as tabulated below.

### Effects of Irradiation on Aflatoxin Production in Foods.

| *S. No.* | *Commodity* | *Irradiation Dose.* | *Effects on Aflatoxin Production* | *Reference* |
|---|---|---|---|---|
| 1. | Groundnut | Increasing Dose | Diseased aflatoxin production | 1 |
| 2. | Ground nuts inoculated with spors of *A-flavus* | 500 karad | Inhibited aflatoxin production | 1 |
| 3. | Wheat | 75 karad | Increased aflatoxin production | 2 |
| 4. | Do | 1, 2.5 and 4 karad | do | 3 |
| 5. | Maize | 75 karad | do | 2 |
| 6. | Sorghun | do | do | do |
| 7. | Pearl millet | do | do | do |
| 8. | Soyabean | Increasing dose | Decreased aflation production | 1 |
| 9. | do | 5.. karad | Inhibited aflatoxin $B_1$ production | 2 |
| 10 | Potato | 10 karad | Increased aflatoxin production | 2 |
| 11. | Onion | do | do | 2 |
| 12. | Red paper | Increasing dose | Decrease aflatoxin production | 4 |
| 13. | Palm juice | do | do | 1 |

Behre et. al also showed aflatoxin production in the irradiated wheat grain (in storage) but which was observed to be lower than in the unirradiated controls. Priyadarshani *et. al.* irradiated wheat at different dose levels from 50 to 250 karads showed as dose dependent susceptibility to aflatoxin production. No correlation was observed between the degree of irradiation and growth of fungus.

Kesavan[7] *et.al.* find that irradiation with $^{60}Co$, γ-rays, induces free radicals, which remain'trapped' in the solid matrices of seeds for considerable lengths of time. Israni[8] *et. al.* on the basis of detailed ESR have established free radicals extended life times upto several weeks even at room temperature. Lower the moisture content of the seeds, longer is the life time of the radiation-induced oxygen-reactive free radicals. Kesavan[9], however, did not arrive at this conclusion due to the fact that the extent of chemical change in the solid dry state is much less than that would occur when substantial amount of moisture is present. When the radiation-induced free radicals trapped" in solid matrices come in contact with water and/or higher temperature (as during cooking) the free radicals rapidly dissipate without causing extensive radiolytic changes in the seeds. Besides, free radical reactions are also common during conventional (thermal) food processing.

Later on the joint FAO/IAEA/WHO Committee concluded that no toxicological clearance would be necessary for food materials exposed upto 10 kGy.

Irradiation of wheat involves only 0.75 kGy. Further, studies on freshly irradiated have no relevance with the food irradiation process which is meant for prolonged storage of wheat and not for immediate feeding. This precludes the presence of polyploid cells or dominant and lethal mutations.

The safety of high-dose irradiated foods has been evaluated in many feeding studies conducted over the past four decades that have involved a variety of laboratory diets and food components given to humans and a broad cross-section of animal species, including rats, mice, dogs, quails, hamsters, chickens, pigs and monkeys. These investigations included sub-acute, chronic, reproductive, multigeneration and carcinogenicity studies, have been conducted under a variety of experimental protocols and have covered a range of doses. Numerous evaluations for mutagenicity have also been made *in vitro* and *in vivo* systems.

The data can be extrapolated to humans by virtue of relevancy of the composite nature of the food materials used and the diet administration manner. Thus available data at present is deemed adequate for evaluating the safety of high-dose irradiated foods for human consumption.

Irradiation conditions are important factors in determining the quality of all irradiated foods as well as in evaluating the potential toxicological effects of irradiating food and feed. For instance, the presence of oxygen during irradiation leads the production of peroxides and the potentially oxidative toxic agents likely to influence the nutritional quality and palatability of the diet. On the other hand, removal of oxygen before irradiation, especially from lipid components in food, limits the production of these compounds like canning/packaging under vacuum or under nitrogen thus limiting the oxygen content in the food. Enzyme inactivation by heating is also helpful because irradiation is not an effective enzyme inactivator.

Dry or dried foods with a lower water content, like spices, can be irradiated at high doses in the presence of oxygen with minimal degradation. Thus proper food irradiation conditions limit the oxygen content and require that the foods, other than dry products, should he irradiated at freezing temperatures. This minimizes undesirable chemical reactions that affect organoleptically. It is imperative that the applied dose must exceed the minimum needed to achieve commercial sterilization.

Production of irradiated foods of wholesome and palatable quality depends upon their method of preparation. As imperative in any food processing procedure the use of high-quality raw materials is a precondition to obtain high quality processed foods. Amount of water, protein, lipid and carbohydrate ascertain the conditions of irradiations, temperature and the atmospheric conditions determine the conditions during irradiation. For sample, unwanted side effects can he minimized by heat-inactivating the proteolytic enzyme, vacuum-packing the food in a can or flexible pouch and irradiating these foods at freezing temperature.

*Source:-* (i) Wierbicki E. *et. al., J. Food Sci.*, 1977, 42, 338-343.
(ii) Wierbicki E.; *Technology of Irradiation Preserved Meats*, Vol.(I), Chicago, American Meat Science Asson, 1980, pp. 194-197.

For evaluation of safety of irradiated foods the focus has been primarily on teratogenic, mutagenic and carcinogenic end points. It is difficult to identify any other food processing technology the safety of which has been supported by so many animal toxicity studies. Following the 1981 Bureau of Irradiated Foods Committee Report, in which it was concluded that, on the basis of studies on the radiation chemistry of foods, an adequate margin of safety can be demonstrated for foods irradiated below 10 kGy and for dry and dehydrated spices that are irradiation sterilized, the FDA reviewed all available animal studies to determine their adequacy and to evaluate the toxicological evidence. This review of 400 studies, resulted in over 250 being "accepted" or "accepted with reservation," and about 150 being "rejected." Some 20 review articles were not categorized. The rejected studies were on the basis of

(i) the radiation was not properly reported, or
(ii) the number of animals per group was not reported, or
(iii) the number of animals per group was small(<5), or
(iv) the study was conducted with controls fed on non-irradiated diet; or
(v) the diet fed was determined to be nutritionally inadequate; or
(vi) the studies were conducted at a laboratory that was considered by the FDA to be violation of good laboratory practice. Nevertheless, their inclusion in the present evaluation provides a broader perspective on the diverse data obtained.

## Sub Chronic Studies

Many sub-chronic studies on safety have been conducted. These studies examined the safety and nutritional adequacy of a variety of dietary items and complete laboratory diets treated with high-dose irradiation. The vast majority of these studies reported no toxic effects in laboratory animals after consumption of high-dose irradiated foods. Other detailed investigations of similar adverse findings in sub-chronic toxicity studies ultimately demonstrated that they were attributable either to pre-existing nutritional deficiencies in the diets or to nutrient degradation not unique to irradiation.

## Carcinogenicity and Chronic Toxicity Studies:

Several studies on high-dose irradiated diets have been conducted. 17 studies were on rats, 3 on mice and 1 on pigs. These are tabulated at the end of this chapter. This data is unique in the assessment of all food related treatments and processes.

No irradiation-related increases in tumors occurred in any of the studies that involved administering high dose irradiated foods or diets to rats/mice. Similarly, no irradiation induced changes in reproductive function were reported in the multigeneration reproduction phases of the combined carcinogenicity-reproduction studies.

Results of chronic toxicity studies conducted on mice, dogs and monkeys. Monsen found that the effects were due to deficiency of iron and copper in the diet. The other chronic studies in mice did not show any adverse effects due to high-dose irradiation diet or to the high-dose irradiated dietary components.

## Reproduction and Teratology Studies

No observable differences were found between rats fed an irradiated diet (50 kGy) and an autoclaved diet (15 min. at 120°C) and those fed a control diet with respect to growth, feed consumption, reproduction, haematology, urinary and organ histo-pathology parameters.

In another study with pigs, no deviations were found in feed consumption, growth, haematological and biochemical parameters in the animals. Another conclusion was that there was no treatment related effects in the growth and reproduction of pigs fed on irradiated/autoclaved feed for three generations. The next study revealed that there were no irradiation treatment at 50 kGy related effects in

(i) feed consumption
(ii) growth
(iii) mortality
(iv) haematology
(v) biochemistry of blood and urine
(vi) organ weights
(vii) histopathology
(viii) tumor incidence and
(ix) the concentrations of nitrosamines in the ham made from the aforesaid irradiated pig meat based ham did not change with added nitrite or irradiation dose.

Read *et.al.* of U.S.A. Medical Research Laboratory have concluded that the variations in reproductive performance does not indicate toxicity. They reported that increased cytochrome oxidase activity is not affected in diets with fruits and vegetables and suggsted that the probable cause of increase was the meat components rather than irradiation.

## Human Clinical Studies

U.S. Army evaluted the wholesomeness of foods treated with high dose irradiation. Subjects consumed irradiated foods for 15 days separated by control and washout intervals. Irradiated foods were tested for sterility and the presence of bacterial exotoxins prior to human consumption.

No toxic effects were observed for any experimental diet, regardless of the proportion of the high-dose irradiation food. No clinical changes were detected in any individual from base line to post-exposure evaluations, or at follow up examinations upto one year post-exposure.

In a second study, 10 volunteers consumed a diet in which 32% of calories were derived from irradiated canned pork treated to 30 kGy. The irradiated canned pork was stored at room temperature for one year prior to consumption; control pork was fresh and obtained locally. There were no adverse clinical effects and no prolongation of prothrombin time for any individual or group following consumption of high dose irradiated pork.

After several studies of foods carried out in the 1950's and 1960's, many of which were processed under conditions that would not he considered as having followed "good irradiation practice," it has been ascertained that irradiated foods using a variety of sources under a variety of conditions are toxicologically safe. The carcinogenicity and mutagenecity studies with irradiated foods and feeds have not demonstrated any treatment related effect.

Based on the body of toxicological data it has been concluded:-

(1) Food irraditation is toxiclogically perhaps the most thoroughly investigated food processing technology.
(2) Animal studies are suitable models and predictions from them are supported by human studies.
(3) The sensitivity of the methods used to assess safety is adequate, and many studies purposely used higher dose and larger amounts of irradiated food in an attempt to elicit a positive response.
(4) The large number of toxicological studies, including carcinogenicity bioassays and multigeneration reproductive toxicology evaluations did not demonstrate any short-term or long-term toxicity related to the process.
(5) With the exception of a few easily rationalised results, the highly diverse and sensitive mutagenicity studies on a variety of foods, including radiation-sterilized chicken, are overwhelmingly negative.
(6) Foods that are appropriately prepared, packaged and irradiated to high doses under proper conditions to sterilize them should be deemed SAFE.

**High-dose irradiation study types - rat studies**

| *Study type* | *Duration* | *Author* |
|---|---|---|
| Subchronic | 90 days | Malhotra & Reber |
| | 8 and 9 weeks | Malhotra & Reber |
| | 8 and 14 weeks | Malhotra, Reber & Norton |
| | 8-12 weeks | Read et al. |
| | 8-12 weeks | Read, Kraybill & Witt |
| | 54 days | McGown, Lewis & Waring |
| | 200 days | Rojo & Fernandez |
| | 84 days | Brin, Ostashever & Kalinsky |
| | 280 days | Lang |
| | 84 days | Verschuuren, van Esch & van Kaay |
| | 90 days | van Logten et al. |
| | 120 days | Metwalli |
| Reproduction | Teratology 15 days | IFIP |
| Carcinogenesis | 2 years | Teply & Kline |
| | 2 years | Bone |
| | 2 years | Teply & Kline |
| | 2.5 years | van Logten et al. |
| Combined carcinogenesis and reproduction | 2 years | Mead & Griffith |
| | 2 years | Teply & Kline |
| | 2 years | Read et al. |
| | 2 years | Blood et al. |
| | 2 years | Becker et al. |
| | 2 years | Richardson |
| | 2 years | Richardson, Ritchey & Rigdon; Rigdon (324) |
| | 2 years | Phillips, Newcomb & Shanklin Tinsley, Bone & Bubl; Bone |

| *Study type* | *Duration* | *Author* |
|---|---|---|
| | 2 years | Barna |
| | Lifetime | Saint-Lebe |
| | 3 generations | |
| | 2 years | Radomski et al. |
| | 3 years | Renner & Reichelt |
| | 2 years | Tinsley, Bone & Bubl |
| | 2 years | Bubl & Butts, Bubl |
| | 2 years | Phillips, Newcomb & Shanklin |
| | 2 years | Aravindakshan et al. |
| | 2 years | Paynter |
| Subchronic | 60 days | Maffei, Mazzali & DeSantis |
| Chronic | 14 months | Teply & Kline |
| | 19 months | Monsen |
| Reproduction | Repro. and teratology 20 days | Raltech Scientific Services |
| | Lifetime, 3 generations | Saint-Lebe |
| | Repro. and teratology 200 days | Porter & Festing |
| Carcinogenesis | 750 days | McKee et al. |
| | | Dixon et al. |
| | 12-28 months | Deichmann |
| | | Radomski et al. |
| | 2 years | Calandra & Kay |
| | 730 days | Raltech Scientific Services |
| Combined carcinogenesis and reproduction | 730 days | Biagini et al. |
| | 300 and 600 days, lifetime | Thompson et al. |
| | Lifetime, 3 generations | Bugyaki et al. |

## REFERENCES


1. Ogbadu, G.; *Microbios*, (1980), 27, 19-26.
2. Priyadassheni, E ànd Tulpule, P.G., *Food Cosmet Toxical*, (1976), 14, 293-5.
3. Niles, E.V., *Trans.Brit.Mycol.Soc.*, (1978), 70, 239-248.
4. Ogbadu, G.; *Microbios* Lett; (1979), 10, 139-142.
5. Behre, A.G., Sharma, A., Padwal desai, S.R. and Nadkarni, G.B.; *J.Food Sci.*, (1978), 43, 1102-1103.
6. Priydarshimi, E. and Tulpule P.G., *Food Cosmet. Toxical*, (1979), 17,505-507.
7. Kesavan, P.C. and Chauhan, P.S., *"DAE - BRNS - Symposium on Nuclear Applications in Agriculture, AH & Food Preservation"*, (1994), March 16-18, New Delhi, p. 44.
8. Israni et al. *Current Science* (1993), 65, 486.
9. Kesavan, P.C. *J. Agric.Biol.*, (1978), 7, 93-98.

# Chapter 10

# RADIATION EFFECTS ON LIVING ORGANISMS

## A. Direct and Indirect Effects

The dose of radiation required to produce a lethal effect varies with the complexity of the organisms. The most complex organisms tend to be the most sensitive to radiation. A dose of several hundred rads is lethal to man but destruction of some micro-organisms requires millions of rads. The following sketch presents a schematic representation of the sensitivity ranges of various living organisms.

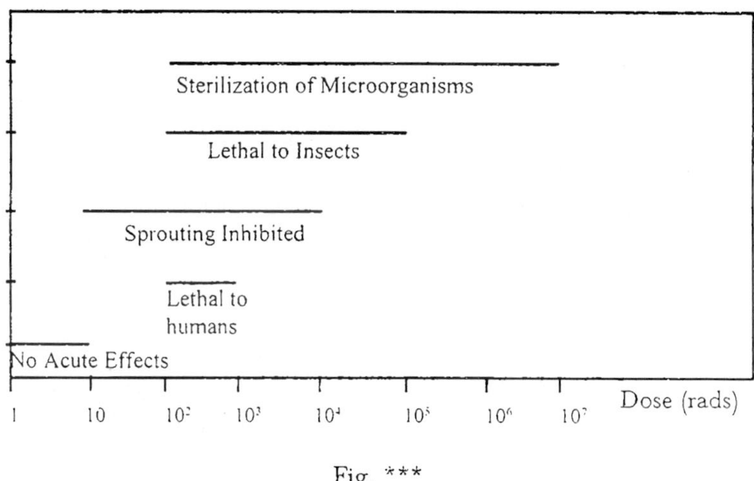


Fig. ***

**Comparison of radiation sensitivity of different organisms.**

There are two basic potential mechanisms by which radiation effects living organisms:

(i) **Direct:** Also known as target mechanism. In it, it is assumed that there is cellular or sub-cellular entity which when hit by a single high-energy particle produces lethality. This theory has applicability to micro-organisms and other unicellular organisms, since the kinetics of their destruction are often in conformity with those predicted by direct action mechanism. This theory, however, does not apply equally well to multi-cellular organisms, unless a single cell or group of cells is regarded as a composite target.

(ii) **Indirect:** These occur via free radicals produced elsewhere than in a primary target or via chemical reactions originating outside the cell even in unicellular organisms.

Radiation effects on cells are complex and some repair of damage as well as occurrence of non-lethal damage (some of which may lead to mutations) must be considered.

## Factors Influencing Radiation Resistance

As with other microbial measures, the response of a microbial cell, and hence its resistance to ionizing radiation depends on:

(i) the nature and amount of direct damage produced,

(ii) the number, nature and lifetime of radiation-generative reactive chemical entities and the inherent ability of the cell either to tolerate radiation damage or to repair it accurately,

(iii) the influence of the intracellular and extra cellular environments on the above factors.

Further ionizing radiation is capable of causing a variety of chemical changes in micro organisms. It is generally assumed that DNA is the most critical targets of ionizing radiation and that the inactivation of micro organisms by ionizing radiation is a result of damage to their DNA.

Ionizing radiation can affect either directly, by energy deposition in this macro molecule, or indirectly, by energy deposition in the surrounding water leading to the formation of diffusive primary radicals, including hydrogen atom ($\dot{H}$), hydroxy radicals ($O\dot{H}$) and solvated electrons ($e_s^-$). The OH radical is the most important; <OḢ> radicals formed in the hydration layer around the DNA molecule are responsible for 90% of the damage; in living cells, the indirect effect is especially significant.

The principle effect included in DNA is chemical alteràtion to the purine and pyrimidine bases and to the deoxyribose component, resulting in a break in phosphodiesterase backbone in one strand of the molecule (single strand break) and, to a lesser extent (5-10%) to breaks in both strands in close proximity (double-strand break). Both prokaryotes (bacteria) and eukaryotes (moulds and yeasts) are capable of repairing many of the different breaks. It is generally believed that microorganisms that are sensitive to radiation cannot repair double-strand breaks, whereas radiation-resistant species have some capacity to do so. Effects on the plasma membrane appear to play an additional role in radiation-induced damage to cells[2].

The major extra cellular environmental factors that influence the survival of irradiated cells are[2,3]

(i) temperature/phase,
(ii) the nature of the gaseous environment,
(iii) water activity,
(iv) pH, and
(v) chemical composition of the food.

These extra-cellular conditions can presumably modify the physical and chemical consequences of intra-cellular deposition of energy. Bacterial spores appear to be less susceptible to modifying factors than are vegetative cells, because of their specific nature.

It is generally recognized that the radiation survival of micro-organisms is not effected appreciably by the rate at which specific dose is absorbed under practical conditions of food irradiation, except where rate of oxygen replenishment is a factor.

1,2,3: WHO Technical Report 890, 1999, p 50-51, WHO, Geneva.

## Lethal effects on Micro-organisms

"Death" of micro-organisms when exposed to radiation can be evaluated by plotting the logarithm of the fraction surviving against dose. ("Death" is in this notation because micro-organisms have the ability of an individual cell to divide and reproduce into a colony of new cells. Thus, a cell which is still capable of some metabolic activities, but not of reproduction has been counted as dead.)

Unlike thermal sterilization, where the effect does not depend solely on the quantity of heat absorbed by the cell but also, in a complex manner, on the intensity factor also temperature on time radiation sterilization is simpler. Here the intensity factor is "dose rate" or the amount of radiation absorbed by the cell per unit time.

Despite "dose rate" has some effects, yet radiation effects are related to dose only as under:

$n = n_0 e^{-D/Do}$, where

n is the number of live organisms after irradiation,

no is the initial number of organisms

D is the dose of radiation received (rad)

Do is a constant dependent on type of organism and environmental factors.

Do i.e. (D zero) can also be deemed that dose of radiation which causes death of 63% of the organisms. $n/n_0$ is linear with dose hence for equal doses the population decreases by a given factor. For working purposes, $D_{10}$ is the constant. $D_{10}$ values of a few micro-organisms are given on page 105. (A)

An important aspect of radiation processing is that even under identical conditions of pH and composition of the suspension media and using the same cultures of micro-organisms, the sensitivity to radiation of different organisms is not necessarily correlated with their sensitivities to heat. The following table on page 105. (B) compares relative sensitivities of different micro-organisms to heating at 121.1$^0$C (250$^0$F) and also to radiation.

## A – Radiation Resistance of Some Micro-organisms

| *S. No.* | *Organism and type* | *Medium in which Irradiated* | $D_{10}$ *(rad)* |
|---|---|---|---|
| 1. | ***Bacterium spores*** | | |
| | *C. botulinum* | | |
| | A36 | Buffer | $0.33 \times 10^6$ |
| | B53 | Buffer | $0.33 \times 10^6$ |
| | B53 | Canned Chicken | $0.37 \times 10^6$ |
| | *C. Sporogenes 367* | Water | $0.22 \times 10^6$ |
| 2. | ***Mold*** | | |
| | *Aspergillus niger* | Saline | $4.7 \times 10^4$ |
| 3. | ***Vegetative bacteria*** | | |
| | *Escherichia coli* | Buffer | $9.0 \times 10^3$ |
| | *Salmonella typhimurium* | Buffer | $2.0 \times 10^4$ |
| | | Egg yolk | $8.0 \times 10^4$ |
| | *Staphylococcus aureus* | Buffer | $2.0 \times 10^4$ |
| 4. | ***Virus*** | | |
| | Organisms responsible for foot and mouth diseases | Frozen solution | $1.3 \times 10^6$ |
| 5. | ***Yeast*** | | |
| | *Saccharomyces cerevisiae* | Saline | $5.0 \times 10^4$ |

## A – Radiation sensitivity and heat sensitivity of different micro-organisms relative to *C. botulinum*) (Type A)

| | *Relative sensitivity** | |
|---|---|---|
| *Organism* | *To heat* | *To radiation* |
| *C. botulinum* | 1.0 | 1.0 |
| *B. coagulans* | 3.0 | 0.5 |
| *B. subtilis* | 0.3 | 5.5 |
| *C. coagulans* | 3.0 | 0.5 |
| *C. coli* | 1.3 | 36 |

*Radiation sensitivity compared on the basis of sensitivity to γ-rays, heat sensitivity to heating at 121°C (250°F).

*Radiation sensitivity is affected by*

(i) Environmental factors during irradiation

(ii) Physiological and genetic differences between strains and cultures of micro-organisms.

The significance of these factors varies as under:

**A**— Temperature has an effect on sterilizing requirements. This effect forms the basis for potential combination processes utilizing both heat and radiation as detailed below:

**Effect of Irradiation Temperature on Radiation sensitivity of *C. botulinum* Irradiated with 9 × $10^5$ rad of $\gamma$-rays.**

| | *Relative number of organisms surviving* | |
|---|---|---|
| *Irradiation Temperature (°C)* | *In Beef Dinner* | *In Buffer Solution* |
| – 196 | 18 | 10 |
| – 100 | 18 | 3.2 |
| – 50 | 18 | 0.05 |
| + 20 | 3 | 0.01 |
| + 60 | 1 | 3 |
| + 90 | 1 | < 0.001 |

The above table reflects a fairly large and complicated effect of temperature on radiation sensitivity of *C. botulinum* in buffer, but a lesser effect in frozen versus unfrozen food. Generally preirradiation sensitizes microorganisms to heat but the effect of temperature on radiation sensitivity is not large. Exceptions are:

(a) Microorganisms are less sensitive in frozen media.

(b) Very high temperatures add their own lethal effects to those of irradiation.

### Effect of Pre Irradiation with γ-rays on Subsequent Sensitivity of Microorganisms to Heat

| *Organism* | *Radiation Dose (rad)* | *Relative No of organisms surviving a given heat treatment* |
|---|---|---|
| *C. sporogenes* | 0 | 100 |
| | $2.79 \times 10^5$ | 40 |
| | $9.3 \times 10^5$ | $5 \times 10^{-4}$ |
| *B. Cerus* | 0 | 100 |
| | $2.79 \times 10^5$ | 12.5 |
| | $5.6 \times 10^5$ | $1 \times 10^{-3}$ |
| *C. botulinum* | 0 | 100 |
| | $6 \times 10^5$ | 1 |

### Radiation sensitivity of some organisms in Frozen state/Unfrozen state

| *Organisms* | *Medium* | *$D_{10}$ dose (rad)* | |
|---|---|---|---|
| | | *Frozen* | *Unfrozen* |
| *B. thermoacidurans* | saline | $1.5 \times 10^5$ | $1.5 \times 10^5$ |
| *E. Coli* | 1M glycerol | $2.7 \times 10^3$ | $10 \times 10^3$ |
| *E. Coli* | Pea Puree | $5.7 \times 10^3$ | $5.7 \times 10^3$ |
| *E. Coli* | Water | $5 \times 10^4$ | $2 \times 10^4$ |
| *S. aurens* | Water | $53 \times 10^3$ | $10 \times 10^3$ |

**B**— The presence of oxygen dose has a significant effect like pH and the type of medium in which the microorganisms are irradiated as shown below.

### Summary of effects of composition of media on radiation resistance of micro organisms

| | | *Organisms* | |
|---|---|---|---|
| *S.No.* | *Factor* | *C. sporogenes* | *B. subtilis* |
| 1. | Effect of chemical composition | More sensitive in saline than in pea puree | More sensitive in pea puree than in saline |
| 2. | Effect of oxygen | Least sensitive in absence of $O_2$ | Least sensitive in absence of $O_2$ |
| 3. | Effect of pH | Least sensitive at pH 7 | No consistent effect |

C– Type of radiation such as electron, γ-rays, x-rays does not have a significant direct effect on dose response.

D– Irradiation in the frozen state reduces radio sensitivity of some organisms as shown in the above table on p. 107. Food should be sterilized in the frozen state to minimize chemical and flavour changes though this needs a greater sterilizing dose.

E– Dose rate (intensity of radiation) doesn't effect sensitivity.

F– Dose rate affects other environmental factors like temperature.

## Radiation effects on Insects

Insects are more sensitive to radiation than any of the microorganisms. Immediate lethality required fairly substantial doses of radiation but reproduction could be prevented by doses of only few thousand rads. The table below gives the mean survival times of insects exposed to γ-rays or to 2.5 MeV electrons from a Van de Graffs electrostatic generator. None of the adult groups which received any of the radiation shown in the ensuring table produced viable eggs. Control (non-irradiated) adults produced several eggs per adult, and over 60% of these eggs developed into larvae.

Eggs from control adults could be prevented from hatching by relatively low doses of radiation as evident from the ensuring table.

### Effect of Cathode Rays on Development of Eggs of *T. Confusum* (Confused Flour Beetle)

### Status Following Irradiation

| *Dose (rad)* | *Day 5* | | *Day 25* | |
|---|---|---|---|---|
| | *Unhatched eggs* | *Live larvae* | *Unhatched eggs* | *Live larvae* |
| Control | 79 | 21 | 36 | 60 |
| 23,250 | 98 | 2 | 98 | 0 |
| 46,500 | 100 | 0 | 100 | 0 |

**N.B.:** Initial count 100 eggs in each group.

## Conclusions

(i) High dose irradiation presents no special microbiological problems. Issues such as selective destruction of micro organisms and potential mutations, which are scrutinized carefully in connection with low- and medium- dose irradiation, are of less concern or are irrelevant at radiation doses higher than 10 kGy.

(ii) Radiation processing of precooked and prepackaged high-moisture food, renders the food shelf-stable and microbiologically safe.

(iii) In establishing food irradiation technologies, the steps of risk assessment for hazardous microbiological agents had already been followed well before the modern concept and terminologies of risk assessment were developed. The four stages of risk assessment have therefore been fully addressed.

   (a) Hazard identification i.e. the most radiation-tolerant pathogenic micro organisms.

   (b) Hazard characterization i.e. toxin formation by *C. botulinum*, the most initial biological agent.

   (c) Exposure assessment i.e. the efficacy of processing for inactivation of spores through the application of the 12D concept, whose very high safety margin takes all reasonable uncertainties into consideration, or the D-equivalency concept in the case of combined antimicrobial agents and/or treatments, and

   (d) Risk characterization i.e. the severity and likelihood of intoxication, which is extremely low for sterilization processes.

Accordingly, the irradiation of foods to doses even above 10 kGy to achieve shelf-stability of high-moisture foods, such as meals and meal components, and to decontaminate low-moisture products, such as spices, herbs and dried vegetables is deemed safe.

From the risk assessment point of view, high-dose irradiation is not different from thermal processing in producing shelf-stable, microbiologically safe foods; both processes have outstanding records of safety.

*Source*: *WHO Technical Report Series (890)*, 1999, p. 76, World Health Organisations WHO, Geneva.

Chapter 11

# RADIATION CHEMISTRY OF IRRADIATED FOODS

## Preamble

The objective of processing with ionizing radiation is to destroy pathogenic and spoilage microorganisms without compromising the safety, nutritional properties and sensory quality of the food. This process, like pasteurization and sterilization, produces physical and chemical changes.

Ionization radiation loses energy to atoms constituting the food. On completion of this physical process there results a formation and reaction of specific chemical entities. These ultimately determine

(i) the destruction of contaminating microorganisms

(ii) the potential formation of a toxic compound and

(iii) the retention of micronutrients, sensory attributes and functionality of the package.

Microorganisms are destroyed primarily because hydroxyl ($OH^-$) radicals formed within their cells react with the base and sugar moieties of DNA. This results in part in breakage of sugar-phosphate bonds and loss of replication function.

A compound capable of eliciting a chronic toxic or genotoxic response can only be formed at a relevant level if a pathway for its formation is possible in principle and competitive in practice.

Micronutrients, particularly vitamins, will be degraded to an extent, depending upon their compatibility against other major constituents for the primary radicals, as well as upon the irradiation conditions, dose level included.

Sensory attributes like flavour, colour, texture will be similarly affected if the constituents normally associated with these attributes can effectively compete for the primary radicals and then follow a reaction pathway that culminates to a stable product with different sensory characteristics.

Package functionality might be affected by the competition between bond-breaking an bond-making reactions. There are influenced by the chemical structure of the material and irradiation conditions.

In short, the effect on the microorganisms, food constituents, packaging material are determined by well- established principles of radiation chemistry.

An understanding of the chemistry involved is especially relevant to the assessment of the safety and applicability of using high-dose irradiation to sterilize foods and render them shelf stable.

The commonality in the microbiological and chemical effects between high and low-dose applications, which primarily involve pasteurization, improved sanitation and enhanced shelf-life is explained by the radiation chemistry. It also provides the rationale for delivering high doses either to dry foods at room temperature or to enzyme inactivated, high-moisture muscle foods at sub-freezing temperatures.

The assessment to be made is in principle, a consideration of the nature and extent of chemical change in the irradiated foods and the impact of these changes would have on the health of the individuals consuming such foods.

If the radiolytic mechanisms by which food constituents undergo certain transformations, the dependence of radiolytic products on absorbed dose, the influence of processing conditions on product yields are all known, it is possible to make a valid extrapolation of the results and the conclusions from one particular food to a class of foods, from one dose regime to another, and from a particular set of condition to an other applicable set.

Taub IA, Angelini P, Merrit C Jr. *"Irradiated food: validity of extrapolating Wholesomeness* data, *J.Fd.Sc.*, 1976,41, 942-4.

## BASIC PRINCIPLES OF RADIATION CHEMISTRY

There is the formation of primary chemical entities in an irradiated matrix. These are ultimately involved in reactions leading to stable radiolysis compounds. These are a consequence of complex physical and physiochemical processes that start with localized interaction of the radiation with the constituent atoms. These continue to the point where these entities are uniformly distributed and react in conformity with the principles of homogeneous kinetics.

*Source:–"Radiation chemistry: principles and applications"*, edited by Farhat Aziz & MAJ Rogers; (1987) – VCH Publishers, N.Y.

Fast-moving, high-energy electrons are generated from x-rays or γ-rays. Their interaction with the atoms when passed through either the photoelectric or Compton process results in the absorption of energy leading to ionization and excitation of constituent molecules. This energy deposition process occurs within $10^{-16}$ seconds, followed by many high-energy processes including energy migration and con-molecule reactions. Many relaxation and thermalization processes take place, including electron salvation; and some reactions occur simultaneously with diffusion away from the site of initial formation. These processes occur within about $10^{-11}$ seconds.

Subsequently the more stable but nevertheless reactive entities in thermal equilibrium with the matrix begin to diffuse and react, primarily with each other, but also with solutes present at high concentration. These further processes which lead to a relatively uniform distribution of radicals, occur within about $10^{-7}$ – $10^{-6}$ seconds. The formation of these stable entities and reactive radicals can be thought of as the "direct effect". Subsequently, the fate of the precursor radicals, the yields of which in pure systems have been generally determined, can be altered through reaction with minor constituents.

The formation of stable radiolysis products through these reactions can be thought of as an "indirect effect." The specific nature of the primary chemical entities formed initially and the precise amount of them that might become uniformly distributed depends on the molecular nature of the matrix.

## Water Radiolysis

Water is a constituent of each food. It contains many dissolved solutes of interest hence its radiolysis is of particular interest.

When water is irradiated, the ionization produces an energetic electron and a cation radical, while excitation produces an excited water molecule. The ejected electron, after losing energy and reaching thermal equilibrium with the surrounding water molecules, can be trapped by a favorable configuration of water molecules to produce a solvated electron, $^{-}e_s$, or can be drawn back to the cation, the ensuing neutralization reaction producing an excited water molecule. The solvated electron is a highly mobile, highly reducing primary entity; it is a precursor of many secondary entities.

The excited water molecule can either lose its excess energy or dissociate into two other primary entities: hydrogen atoms ($H^{-}$) and hydroxyl radicals ($\dot{O}H$), both are highly mobile, the former being a strong reductant, the later a strong oxidant. The cation radical is extremely short-lived, its major pathway for reaction being proton transfer to water, producing a hydronium ion $(H_3O)^+$ and $\dot{O}H$. Various recombination and cross-combination reactions of the primary radicals are possible, the combination of $\dot{H}$ and $\dot{O}H$ regenerating water ($H_2O$). Such reactions occur simultaneously with diffusion away from the site of energy deposition, which results in a specific number of primary radicals being distributed through out the medium.

## Irradiation Parameter Effects

### *(i) Atmosphere*

The presence or absence of oxygen ($O_2$) in the head space can influence the chemistry by introducing new pathways for reaction. By virtue of its high electron affinity, $O_2$ reacts with $e^-_S$ and $\dot{H}$ and with organic radicals. Reaction with the former leads to the formation of $\bar{O}_2$ or $H\dot{O}_2$, which react primarily to yield $H_2O_2$. Reaction with organic radicals leads to the formation of $RO_2^{\circ}$ radicals, which tend to react biomolecularly to form peroxides, but can also decompose unimolecularly to $\dot{R}$+$H\dot{O}_2$. This pathway is particularly relevant to the reaction in which a hydroxyl radical adds to an aromatic ring, forming a further radical, in the presence of oxygen. This results in simple hydroxylation, as in the conversion of phenylalanine to tyrosine. Reactions with lipid radicals can involve subsequent hydrogen abstraction, resulting in another radical and $H_2O_2$.

***(ii) Temperature***

Because many radical reactions proceed with very low activation energies, changes in temperature only slightly increase or decrease the rate constants. However, if there are two competing reaction pathways with different activation energies, the temperature may influence the direction taken, depending on the extent to which it affects the rate constant.

Molecular relaxation and diffusion processes in solids can be influenced by temperature. The extent of certain reactions in frozen aqueous solutions can be strongly temperature dependent . For example, the formation of $NO_2$ by the reaction of $e^-_S$ with $NO_3$ in frozen solutions irradiated at temperatures of –100 to –10°C ,there is only a small increase in yield with increasing temperature until about -30°C, but thereafter it rises sharply with temperature, the ratio of the G value at –10°C to the G-value at –30°C being about 10. The same effect is seen with other indices of reaction like reduction of brown ferrimyoglobin by $e^-_S$ to the red ferromyoglobin. Here too, the rationale for irradiating frozen food starting at a temperature of –40°C can he understood.

***(iii) Phase***

Phase changes can have a substantial influence on the outcome of the radiolysis, primarily as a result of changes in the mobility of the constituent molecules and of any reactive entities derived from them.

Phase affects the formation and distribution of primary radicals as well as their subsequent reactions with and within the confining matrix. For instance, in frozen water the yield of primary radicals is substantially reduced.

*Source*– Taub, I.A., Eiben K., *J. Chemical Physics,* (1968), 49 ; 24 99-2513

These radicals tend to react either with each other or with major constituents in their proximity instead of the with low concentration of solutes that–are constrained from diffusing and are separated by long inter molecular distances.

Taub *et.al* have graphed these histograms below for illustrative purposes.

**Comparison of G-values resulting from irradiation in liquid solutions at 20°C and in frozen solutions at –40°C; the abscissa indicates the formation (+) or loss (-) of the indicated analyte upon reaction of different radicals with specific solutes[a]**

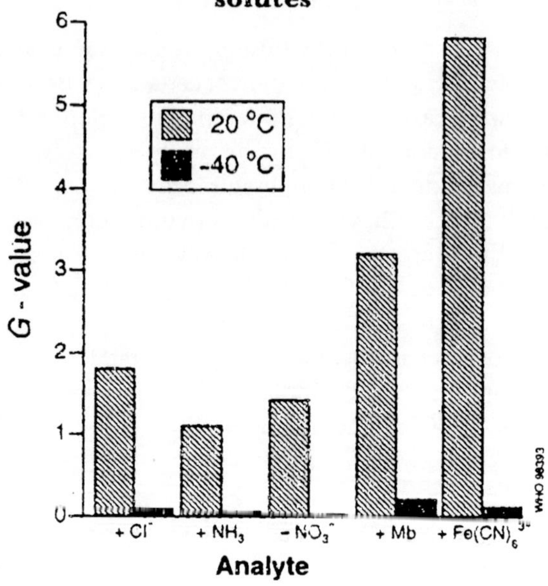

*Source:*– Traub I.A; Kaprielian RA; Halliday, J.W, *Proceedings, International Symposium on Food Preservation by Irradiation*; Vienna, IAEA, (1978), Vol I , 371-384.

The first four histograms from the left correspond to reactions of $e_S^{-}\cdot^{-}$ with solutes, including the $NO_3^{+}$, the histogram on the right pertains to the oxidation of the ferrocyanide ion by $\dot{OH}$, whose yield is doubled by the reaction of $e^-_S$ with $N_2O$. The figure also illustrates the rationale for using frozen (or dry solid) foods when applying high sterilization doses of irradiation. It also helps to explain why the extent of chemical changes in such foods is quantitatively not unlike that observed in chilled foods receiving lower pasteurizing irradiation doses:

## Dose Rate

The rate at which energy is deposited influences the rate of increase with the concentration of reactive radicals, which could have an influence on their reaction pathways.

*Source:*– Taub, I.A, *Preservation of Food by Ionizing Radiation*, Vol. 2, edited by Josephson, ES, Peterson M; Boca Raton, FL CRC Press, 1983, pp. 125-166.

If there is the possibility of competing bimolecular and unimolecular (or pseudo –unimolecular) reactions, then the bimolecular process is favoured at high radical concentrations. Accordingly in the following competition between bimolecular combination and unimolecular dissociation of an acyl radical, the likelihood

$$2RCH_2CO \rightarrow RCH_2COCOCH_2R \quad (a)$$

$$RCH_2CO \rightarrow RCH_2 + CO \quad (b)$$

of reaction (a) increases with a substantial increase in dose rate, since the concentration of $RCH_2CO$ will be higher at any one time.

## Dose Dependence

Generally the yield of a particular radiolysis product will increase linearly with dose, exceptions are there depending on variability of the dose range employed.

If the precursor of a particular product is a minor constituent in the food, then the yield of that product will increase till the depletion of the precursor. There after it shall be constant, unless it is reactive towards primary entities. A product of a minor constituent which has the potential of competing for primary entities will ultimately get depleted and produce a secondary product. Since energy deposition is partitioned according to the mass fraction of the components, major radiolysis products are expected to be derived from major constituents–water, proteins, lipids and carbohydrates – and to be formed in yields that are linearly dependent on dose in the practical range anticipated for radiation sterilization.

On the other hard, if the precursor of the product is a major food constituent and the dose is insufficient to result in a product concentration capable of competing for primary chemical entities, then the yield of that initial product will remain linearly dependent on dose. At high doses, however some products with high reactivities will reach high enough concentrations to compete, the result will be that their yield remaining constant and the yields of secondary products derived from them increasing linearly with increasing dose.

## MAJOR CONSTITUENTS

The radiolysis of major constituents in as complex a matrix as a muscle food can be understood by considering each constituent separately. The reason being that the chemistry tends to be compartmentalized. In chilled and especially, in frozen muscle foods, the deposition of energy and the resultant chemical reactions occur over short ranges within almost distinct and immiscible phases.

The main constituent, water (65%), surrounds and to differing extents suffuses the other two major constituents, proteins (20%) and lipids (15%). The water extensively suffuses the proteinaceous myofibrils comprising primarily myosin and actin. The depot fat comprising different triglycerides is essentially separate. There are many interfaces between the fact and other constituents but interfacial reactions are not deemed significant.

Each constituent phase contains soluble materials. Water contains sarcoplasmic proteins including myoglobin and albumin, as well as diverse vitamins, salts and small peptides; the proteins can bind certain compounds, including thiamine; and the fat contains vitamins A and E, as well as other compounds.

Carbohydrates though a small fraction of muscle tissue, they are important dietary components and constitute a large fraction of other foods (e.g. vegetables and cereals) that might be irradiated together with fats & proteins, The basic radiolysis of the proteins, lipids and carbohydrates constituting the macro nutrients of muscle foods have been considered below within this frame work.

## Proteins

The major effect of the ionization and excitation processes accompanying energy deposition in proteins is the formation of a peptide back bone radical, corresponding to scission of the back bone C-H bond as observed in the electron spin resonance (ESR) spectrum obtained upon irradiating a suspension of myosin/ actomyosin. A broad asymmetric doublet is seen. Spectral analysis, based on spectra obtained by irradiating diverse peptides, indicates that the doublet is a composite of many spectra in which the unpaired electron interacts primarily with a single proton bound to the C-atom linking the side chain moieties of constituent amino acids to the peptide back bone. The spectrum also shows a much less intense contribution of radicals corresponding to $\dot{H}$. addition to the benzene ring of aromatic amino acid moieties. The ESR spectrum produced by irradiating the suspension of the myofibril bundles on the whole muscle is the same. This indicates that the aggregation of myosin into more complex structures does not effect the mechanism of radical formation, which occurs on a molecular scale.

Despite the radiolysis of proteins and water being analogue the formation of peptide radicals and other reactive entities and the pathways for their subsequent reaction are complex. However, electron attachment to the peptide carbonyl group proceeds via preferred steps common to all proteins.

The first step leads to an observable carbonyl anion radical that dissociates into a stable amide and an alkyl radical; another step involves the abstraction of hydrogen from the C-H back bone by this radical to form a stable compound and the peptide linkage. The last step is the bimolecular reaction of this radical, either to dimerize, forming a cross linked myosin, or to disproportionate, reforming the myosin and forming an imine with its hydrolysable N=C linkage.

Despite the size of the peptide radical, it will react at –10°C over the course of hours, upon thawing it will disappear rapidly.

Sequences of reactions initiated by $\dot{H}$ and $O\dot{H}$ will differ from that initiated by the solvated electron but some of same intermediates will be formed. $\dot{H}$ could react at the back bone C-H and at the carbonyl group, but is more likely to add to aromatic or heterocyclic rings in the side chain moieties.

By implication, proteins in irradiated frozen meats would get slightly altered by some aggregation and fragmentation, to an extent limited by the low G-values for primary radical formation. Though there would be only slight discrimination among the amino acids moieties affected. This has been corroborated through electrophoresis to assess changes in protein molecular size; enzymatic hydrolysis to assess digestibility and amino acid analysis of acid-hydrolyzed samples; particularly there is no dose-related change in amino acid composition and the dose range anticipated for use in sterilization. Albeit within the limits of sensitivity for analyzing amino acids.

***Metaloproteins***:– The presence within a protein molecule of a metal ion that can be oxidized or reduced provides additional pathways for reactions of both primary and secondary entities. This potential for modifying the radiation chemistry is especially significant for small or globular proteins, such as the pigment myoglobin. Because the metal ions represent such a small proportion of the total host protein mass, and are fixed in specific location within the molecular geometry, they can only influence the reaction of primary entities or small secondary radicals when at exposed and accessible sites, and their influence on the fate of the radicals formed in the host protein is limited to relatively short ranged reactions. Nevertheless, the reactions are distinctive and have been studied in detail in both model and food systems.

*Source*:- Shieh J.J. and Whitburn K.D. *et.al.* quoted in *WHO Tech. Report series 890*, (1999), pp. 20-23. WHO, Geneva.

The basic radiolysis of each macronutrient of muscle foods is explained in the ensuing pages.

(i) **Lipids** : The major effects of the ionization and excitation of the lipids triglycerides in foods are:–

(a) disruption of the bond between the fatty acid and glycerol portion,

(b) formation of dominant triglycride radical corresponding to an unpaired electron on the carbon atom in the $\alpha$–positive relative to CO group,

(c) stable products are derived from this radical,

(d) In triglycrides with polyunsaturated fatty acid portions, radicals would be formed by splitting of a C-H bond near the unsaturated functional group that H is lost from the weak C-H bond $-CH_3$ group in the linoleic moiety,

(e) The molecular processes are independent of the way in which the triglycerides are organized, so the same radicals are observed in isolated systems and in a complex muscle food,

(f) Sequence of reactions lead to the most stable radicals in lipids, viz. in tripalmitin by electron reaction, the first effect is electron attachment to the $CO^+$ radical. This radical then dissociates into the stable palmitic acid anion and an alkyl radical which in turn abstracts a hydrogen from the carbon alpha to the CO group on another tripalmitin molecule forming the main radical $RCH_2O\ (C=O)C^0H\ (CH_2)_{15}\ CH_3$. This large, slowly diffusing radical reacts either by combination forming a stable dimer, or by disproportionation, reforming the original tripalmitin and forming an unstatured analogue as sketched below.

$$RCH_2O\,(C{=}O)\,CH_2(CH_2)_{15}CH_3$$

$$\downarrow e^-$$

$$RCH_2O\,(\dot{C}\,O^-)\,CH_2(CH_2)_{15}CH_3$$

$$\downarrow$$

$$R\dot{C}H_2 + (^-O_2C)\,CH_2(CH_2)_{15}CH_3$$

$$RCH_2\,O\,(C{=}O)\,CH_2(CH_2)_{15}\,CH_3$$

$$RCH_2O\,(C{=}O)\,CH\,(CH_2)_{15}\,CH_3 \qquad RCH_2O\,(C{=}O)\,CH_2(CH_2)_{15}\,CH_3$$

$$| \qquad\qquad\qquad\qquad + $$

$$RCH_2O\,(C{=}O)\,CH\,(CH_2)_{15}\,CH_3 \qquad RCH_2O\,(C{=}O)\,CH{=}CH\,(CH_2)_{14}\,CH$$

By inference, same reactions will take place and a similar distribution of products corresponding to the constituent fatty acid composition. For example, a sequential formation and conversion of radicals is observed in irradiated beef fat leading to the preferred triglyceride radical at $\dot{O}C$. The net chemical change will be small in products like fatty acid, (R-COOH), hydrogen and propanediol-esters (derived from the initial alkyl radical) predominating. Volatile compounds yield though very low yet are determined. These volatiles provide very useful insights into the common pathways for reaction.

Notwithstanding, cyclic compounds like alkylcyclo butanones though not found in non-irradiated lipids yet are produced and developed in lipids subjected to high temperature making their residual concentration quite low. These cyclobutanones appear to be specific to the irradiated process yet the above facts need to be borne in mind.

***(ii) Proteins*** : Ionization and excitation processes when deposit energy in proteins, the major effect is the formation of a peptide backbone radical, corresponding to split of the backbone C-H bond. Detailed spectral analysis has revealed that the doublet is a composite of many spectra in which the unpaired electron interacts primarily with a single proton bound to C-atom linking the side chain moieties of constituent amino acids to the peptide backbone. The contribution of radicals corresponding to $\dot{H}$ addition to the benzene ring of the aromatic amino acid split-halves is quite mild. When the myofibril bundles or the whole muscle suspension is irradiated then the aggregation of myosin into more complex structures does not affect the mechanism of radical formation, occurring on a molecular scale.

Radiolysis of proteins and water is similar, however, in the former the formation of the peptide radicals and other reactive entities are more complex. Here the first step leads to an observable carbonyl anion that dissociates into a stable amide and an alkyl radical; another step involves the abstraction of hydrogen from the C-H backbone of this radical to form a stable compound and the peptide radical; the last step is the bimolecular reaction of this radical, either to dimerize, forming a cross-linked myosin or reforming the myosin and forming an immune with its hydrolysable N=C linkage. Despite the size of the peptide radical, it will react at $-10^\circ$C over the course of hours.

Sequences of reactions initiated by $\dot{H}$ and $O\dot{H}$ will differ from that initiated by the solvated electron, though some of the same intermediates will be formed. It implies that proteins in irradiated frozen meats would be slightly altered by some aggregation and fragmentation. Further, there would be only slight discrimination among the amino acid moieties affected and there is no dose-related change in amino acid composition over the dose range anticipated for use in sterilization.

***(iii)Carbohydrates :*** Direct ionization and excitation of carbohydrate molecules, may these be polysaccharides, disaccharides or monosaccharides, results in the breaking of C-H bonds and the disruption of ether linkages.

In solids such as starches (polysaccharides), bond breakage is mainly at the glucosidic linkage, leading to depolymerisation and eventually to radicals centered on the C1 and C6 positions. Radicals formed in starches of different origin are identical. The results are subjectively the same when the irradiation is carried out with or without oxygen present and at either room temperature or at 77K.

The results for radiolysis products formed in starches and their oligomers are also the same. However, the quantities formed are a function of dose up to a value depending on the product concerned.

In solution the resulting radical sites would be at all the carbons, with some preferences for $C_1$ and $C_6$ positions. In may ways, there reactivity of $e^-_S$, $\dot{H}$, $O\dot{H}$ towards saccharides is very much like that towards alcohols or simple ethers.

In short, the implication of carbohydrates radiolysis for radiation sterilization of muscle foods is that the chemical consequences will be minor except where sucrose is purposely added, the level of carbohydrate in the tissue being about 0.5%. At low concentrations and with relatively low reactivity, carbohydrates are not likely to compete for the primary radicals. Predictability of radiolysis of sugar in fruits is substantially different from radiolysis of concentrated sugar solutions.

***(iv) Vitamins*** : Like thermal treatments, radiation processing of foods causes some loss of vitamins. Some vitamins are immune to ionizing radiation while the rest are radiation sensitive. As is well known, the entire clan of vitamins consists of either fat-soluble or water-soluble vitamins. Among the fat soluble vitamins, the gradation of radiation sensitivity is Vit. E > Carotene > Vit. A > Vit. D > Vit. K. Among the water soluble vitamins the position is Thiamine ($B_1$) > Vit. C > Vit. $B_6$; Vit. $B_6$ > Folate, niacin (nicotinic acid) > Vit. $B_{12}$.

Several factors influence the vitamin's radiation resistance. These factors are (i) food composition, (ii) packaging atmosphere, (iii) temperature during irradiation, (iv) post-irradiation storage temperature, (v) the presence/absence of oxygen in the packaging atmosphere.

Ascorbic acid, by virtue of its carbonyl group and a double bond is highly reactive to $e^-_S$, $\dot{H}$ and $O\dot{H}$. It is reduced to an intermediate radical by $e^-_S$ and $\dot{H}$ and is oxidised to the relatively stable tricarbonyl radical ion by $O\dot{H}$, which in turn is involved in biochemical processes not initiated by irradiation. Its most likely reaction pathway is a complex disproportionation reaction that regenerates ascorbic acid as well as produces dehydroascorbic acid, still exhibiting vitamin activity.

Experience has shown that Vit. C rich foods like fresh fruits and fruit juices, vegetables and potatoes are generally unsuitable for high dose irradiation. Their sensory qualities are affected which are undesirable. Consequently-the average dose of irradiation (kGy) for onions is only 0.06, potatoes 0.10, mango, raisins, figs and dried dates 0.50, ginger, garlic and shallots (small onion) is only 0.09.

Summing up, most of the foods providing Vit. A in the human diet are milk and its products like butter and cheese. Vit. C from vegetables and fruits. These are not allowed to be irradiated at more than 1 kGy hence do not suffer vitamin losses. The major sources of Vit. E in human nutrition are margarine, butter and vegetable oils and fats. These foods do not require irradiation due to absence of microbiological problems. Further loss of Vit. E gets prevented indirectly because these high fat foods develop undesirable sensory quality when irradiated to high doses, which in turn is restricted to only a maximum dose of 4.0 kGy for meat and meat products including chicken. And the water soluble vitamins, the data put up by various research works and research groups, demonstrate once again the improved retention of thiamine when irradiation is carried out at lower temperature with electron irradiation than with $\gamma$-irradiation. No anti-metabolites have been formed to thiamine and pyridoxine in irradiated chicken and beef so also is the absence of anti vitamin factors in any of the meats tested.

No significant loss of riboflavin has been noted in cheddar and mozzarella cheeses, yoghurt bars, ice cream, and non-fat dry milk sterilized with a dose of 40 kGy at –78°C in a nitrogen atmosphere. *These conditions are not commercially available in India which perhaps has kept milk and above dairy products away from the list of foods permitted for irradiation.*

***(v) Salts :*** Except nitrates and nitrites most of the common salt additives to foods are relatively inert towards the primary radicals formed in water. Nitrate is highly reactive in solution towards $e^-_S$. The mechanism of reaction leading to nitrate formation is :

$$e_s^- + NO_3^- \rightarrow NO_3^{2-} + H+ \rightarrow NO_2 + OH^+$$

$$2NO_2 + H_2O \rightarrow NO_2^- + NO_3^- + 2H^+$$

The overall stoichiometry is

$$2e_s^- + NO_3^- + H_2O \rightarrow NO_2^- + 2OH^-$$

which means that one mole of nitrate is formed for every two mols of electrons reacting with a mole of nitrate.

So far radiation sterilization of bacon or ham is concerned, the likelihood of reactions $e_s^- + NO_3^- \rightarrow NO_3^{2-}$ is low. Besides the frozen system, the low levels of $NO_2^-$ & $NO_3^-$ would have to compete with high levels of constituents that are reactive towards the electron.

***(vi) Nucleic Acids :*** These acids are factually a very small fraction of the food mass yet their radiolysis has a bearing on the microbial destruction. Notwithstanding, it is unlikely that any altered bases in the food could be incorporated into human DNA. Its synthesis involves enzymes that act on precursors of the base, not on the bases themselves, so competition between normal altered bases is not a factor. Moreover, even if an altered base has entered for incorporation, the DNA polymerases would remove by cutting any incorrectly matched base.

***(vii) Chemiclearance :*** Unless one knows thoroughly radiation chemistry he is incapable of considering the generic clearance of irradiated foods. The operative principles are :

(i) When foods of similar composition are similarly irradiated their chemical and microbiological responses are similar and they are accordingly, toxicologically equivalent.

(ii) When an irradiated food in a class of similar foods is cleared as safe and adequate for consumption, then others members of that class are, correspondingly deemed wholesome. These principles are reflected in the commonality of intermediates as well as of lipid-derived volatile products as detailed here.

***Commonality of Intermediates :*** Irrespective of the nature and condition of the muscle foods, the same type and behaviour of (a) protein-derived and (b) lipid-derived radicals are observed in all of them upon irradiation. This has been confirmed by using ESR (electron spin resonance) techniques to detect radicals and chromatographic analysis to quantify the yields of the products. This common pattern indicates that the process by which the radicals stable at this temperature are formed are all similar. The subsequent behaviour of the protein radical upon throwing, leading to some aggregation and degradation must also be similar, since gel electrophoretic patterns of the extracted proteins are all similar.

Similarly the commonality in the triglyceride radicals were confirmed by spectral analysis of fats from the meats which were similarly but individually irradiated. The small differences among them being attributable to differences in the triglycerides in the meats (chicken, beef and pork)

## Commonality of Lipid- Derived Volatile Products

The likely hood of C-C or C-O bond scission in the fatty acid moieties bound to the glycerol structure in a triglyceride is significantly smaller. Such scission processes lead to different alkyl, acyl and acyloxy radicals. The alkyl radicals can form stable components by abstraction, by combination to form dimeric compounds, and by disproportion to form two hydrocarbons, one with a double bond of the terminal end

(i) $CH_3(CH_2)_{15}\dot{C}H_2 + RCH_2O\ (C = O)\ CH_2\ (CH_2)_{15}\ CH_3 \rightarrow$
$CH_3(CH_2)_{15}CH_3 + RCH_2O\ (C = O)\ \dot{C}H(CH_2)_{15}CH_3$

(ii) $2CH_3(CH_2)_{15}\dot{C}H_2 \rightarrow CH_3(CH_2)_{15}\ CH_2CH_2(CH_2)_{15}CH_3$

(iii) $2CH_3(CH_2)_{15}\dot{C}H_2 \rightarrow CH_3(CH_2)_{15}CH_3 + CH_3(CH_2)_{14}\ CH{=}CH_2$

It is from types of radical reactions that evidence for commonality and predictability can be obtained.

## Dependence on total fat

In the case of C-C bond scission in the fatty acid chain, the resultant products with about 6 carbons or less will be the same irrespective of the particular fatty acid. Consequently, the yield of pentane, hexane and even heptanes and octane should be closely related to the total amount of fat in the sample. This prediction is confirmed by comparing as a function of fat content the yields of such hydrocarbons from enzyme-inactivated ham, chicken, pork and beef irradiated over a range of dose.

### Dependence on fatty acids

Since C-C bond scission at the $\alpha^-$ and $\beta^-$ portions relative to the carbonyl group in the fatty acid moieties can occur, the subsequent abstraction reaction will result in hydrocarbons with 1 or 2 from C-atoms ($C_{n-1}$ and $C_{n-2}$), respectively. Similarly the alternative disproportionation reaction will result in $C_{n-1}$ and $C_{n-2}$ hydrocarbons with an added double bond. The yield of certain volatile hydrocarbons will depend upon the level of the precursor fatty acid in the triglyceride of the foods being irradiated. For example the yield of heptadecadiene ($C_{17:2}$) normalized /gm of fat /10 kGy of dose is linearly dependent on the proportion of linoleic acid in the fat. This relationship also holds for uncooked products irradiated in the chilled state over a lower dose range. The yield of hexadecatriene ($C_{16:3}$) is also instructive, because it is formed in part when the $C_{n-2}$ radical from linoleic acid reacts by disproportionation and acquires a terminal double bond. Analyses of $C_{16:3}$ from 5 different uncooked products irradiated in the chilled state show that the normalized yields are linearly dependent on dose.

Such similarities in the G-values implies that the formation of the radical and its subsequent reactions are essentially independent of the molecular environment in which the precursor fatty acid moiety exists.

### Dependence an triglycerides

Since the major fate of electrons formed in the ionization process is to react by dissociative attachment to the (C=O) group in any fatty acid moiety of the constituent triglycerides an equal number of the stable fatty acid anions and propanedioldiester radicals will he formed. To emphasize the positional differences of the fatty acid moieties, there radicals are denoted here as $H_2C$ $(O_2R'')$ $CH$ $(O_2R')$ $\dot{C}H_2$, showing the loss of $\bar{O}_2R$ from the 1-position. Upon extracting a hydrogen from other triglycerides, they become stable propanedioldiester products, $H_2C$ $(O_2R'')$ $CH$ $(O_2R')$ $CH_3$. Accordingly, if scission is equally likely from corresponding dioldiester isomers, namely $H_2C$ $(O_2R'')$ $CH$ $(O_2R')$ $CH_3$, $H_2C$ $(O_2R'')$ $CH_2$ $CH_2(O_2R')$ or $H_3C$ $CH$ $(O_2R)$ $CH_2$ $(O_2R)$ will correlate with the level of precursor triglycerides containing the relevant $O_2R''$, $O_2R'$ and $O_2R$ fatty acids.

The adopted procedure for this study was that non-volatile propanedioldiesters were isolated from enzyme-inactivated chicken, beef, pork and ham samples irradiated to 30,60 and 90 kGy. Propane dioldipalmitate and propanediol palmitateoleate yields were found to increase linearly with dose, the slopes being different in each product. This implies that the formation and reaction of the relevant radicals is not specialty sensitive to the specific molecular environment of the precursor triglycerides. It confirms the commonality in the chemistry among diverse triglycerides and the prediction of shifts in product distribution based on a knowledge of compositional differences.

## General Implications

The detailed studies of the type described above about normalized yields of radiolysis products for different products contain enough evidence to substantiate the stated principles. These can be applied to different food classes. The many studies on proteins in diverse foods irradiated to low and high doses, on volatile and non-volatile products derived from fatty acids, their esters and oils, are consistent in their chemical behaviour/chemistry. Results similar to those for proteins and lipids have also been obtained with diverse starches and glucose oligomers. *i.e.* the radicals formed in cereals are the same as those formed in pure starches. All have the same ESR spectral characteristics and the same decay dependence or water content and storage time.

It means that variations in the food matrix containing these same constituents would not alter significantly the course of reactions described earlier and consequently would not affect safety. It is important to stress that whenever high-dose irradiation of meats is carried out, these meats must be in a frozen state and in an oxygen free area. Absence of oxygen significantly reduces the formation of undesirable oxidation products (from lipids) and avoids the loss of certain flavour compounds (in spices).

## Conclusions

The knowledge of what can and does occur chemically in high-dose irradiation foods, which derives from over 70 years of research on radiation chemistry and from over 50 years of research on the radiolysis of food, justifies the following conclusions:

1. Reactions initiated by the irradiation process follow pathways for each major constituent that are predictable and that depend on processing conditions.
2. Overall chemical change, as reflected either in the formation of a stable compound or the loss of a particular constituent, is quantifiable and relatively minor, requiring sensitive techniques to discern that a product had been irradiated.
3. Yields of any product derived from a major constituent will depend linearly on dose, but yields from a minor constituent could remain constant or even decrease once the dose corresponding to the depletion of that constituent is reached.
4. As a consequence of the penetrating power of the radiation permitted for use and of the associated energy deposition process, the yield of products formed or lost throughout the irradiated food will be relatively uniform, varying by less about ± 25%.
5. As a consequence primarily of the effect of phase, irradiating moist foods while frozen and in the absence of oxygen significantly decreases the overall chemical yields by about 80%, so the cumulative effects of irradiating to a dose of 50 kGy at –30°C is essentially equivalent to a dose of 10 kGy at room or chilled temperatures.
6. Compounds found in irradiated model systems that are either far different in composition from the foods of interest or have

been irradiated under extreme conditions do not validly reflect the chemistry (or toxicology) of actual foods, because competitive reactions will occur in the latter that make the formation of such compounds very unlikely.

7. Virtually all of the radiolysis products found in high-dose irradiated foods to date are either naturally present in foods or produced in thermally processed foods, a radiolysis product being defined as a compound that originates from a food constituent during irradiation and that, at least initially, increases in yield with increasing dose.
8. This understanding of the radiation chemistry of foods is vital in assessing wholesomeness.
9. The commonality in the chemistry among the major protein, lipid and starch constituents, with minor chemical differences being accounted for by the slight differences in the composition of these constituents, justifies use of the chemiclearance approach for granting broadly based, generic approvals of high-dose irradiated foods.

Chapter 12

# EFFECTS OF IRRADIATION ON INDIVIDUAL FOODS

All food preservation operations have an element of compromise between maximizing the efficiency of the process and maintenance of optimum product quality. In the case of irradiation there are undesirable side effects which can cause impaired flavor, texture and appearance. Like in other processes the side effects can be minimised and in same cases even eliminated, by changes in process conditions and the seriousness of the problem varies with different foods as shown below.

**Quality Defects Produced by Side Effects of Ionising Radiation and Methods of Preservation.**

| *S. No.* | *Food* | *Defects* | *Dose at which the side effects become important (rad)* | *Method (s) or Prevention* |
|---|---|---|---|---|
| 1. | Beef | off flavour | ~ $2 \times 10^6$ | Irradiation in frozen state |
| 2. | Fish | off flavour | ~ $x\ 10^6$ | Irradiation in frozen state |
| 3. | Flour | poor baking performance | ~ $x\ 10^6$ | Irradiation in frozen state |
| 4. | Fruits | Softening | $<10^6$ | Dose reduction |
| 5. | Milk | off flavour | $<10^5$ | Distillation of volatile off flavour |

| *S. No.* | *Food* | *Defects* | *Dose at which the side effects became important (rad)* | *Method (s) or Prevention* |
|---|---|---|---|---|
| 6. | Pork | off flavour | ~ x $10^6$ | Irradiation in frozen state |
| 7. | Paultry | off flavour, pink discoloration | ~ 3 x $10^6$ | Irradiation in frozen state |
| 8. | Vegetables | Softening discoloration | <$10^6$ | Dose reduction |

In general off flavours are the major organoleptic problem in meats, fish and poultry, probably due to oxidation of fat as well as break down of proteins may be involved. Beef is considered to be more sensitive; pork and poultry the least sensitive Pigment changes also occur. Red pigments of raw meat turning brown and cooked meat and poultry pinkish; some softening of texture also occurs.

In the case of fresh fruits and vegetables, textural changes are important as well as off flavour, and undesirables textural side effects occur at doses $10^5$ rad and above.

Some foods, notably dairy products, are particularly sensitive to radiation induced flavours. Milk for instance, acquires an unpleasant off flavour at doses as low as 10, 000- 500000 rad and some type of chocolates acquire a detectable off flavour at doses below 5000 rad.

The undesirable effects as eliminated entirely by changes in process conditions. Two major approaches are possible, each with several variations.

(i) Reduce the required radiation dose. This can be achieved by using radiation in combination with heat treatment or by using radiation together with preservatives.

(ii) Use the full sterilization dose but to reduce the severity of side effects of irradiation the following techniques are adopted.

(a) The product under irradiation should be in a frozen state.

(b) In an inert atmosphere or

(c) By adding free-radical scavengers like ascorbic acid.

Irradiation in the frozen state increases the dose required for sterilization but this is outweighed entirely by the almost complete elimination of intolerable off flavours as is demonstrated by the following table.

**Effect of Irradiation temperature on Quality of Irradiated beef**

(Dose = 4.5 x $10^6$ to 5.6 x $10^6$ rad)

| *Temperature of Irradiation (°C)* | *Intensity of Defects* | | |
|---|---|---|---|
| | *Flavour* | *Colour* | *Texture* |
| 10 | 4.1 | 3.9 | 2.9 |
| -8° | 2.1 | 2.3 | 2.0 |
| -18° | 1.5 | 2.3 | 1.6 |
| Control | 1.0 | 1.2 | 1.6 |

**NB.** 1= no defect; 9= extreme defect.

According to Karell, typical dose requirement and condition of irradiation are as under. These requirement are based on reducing a potential population of *C. botulinum* spore to $10^{-12}$ of the original population

Sterilizing Dose for following Foods.

| *Food* | *Irradiation Tamp. (°C)* | *Dose (rad) (x$10^6$)* |
|---|---|---|
| Bacon | 15 | 2.3 |
| Beef | -30 | 4.7 |
| Beef | -80 | 5.7 |
| Ham | 15 | 2.9 |
| Ham | -30 | 3.7 |
| Pork | 15 | 4.6 |
| Pork | -30 | 5.1 |

Irradiation is not a particularly effective method of destroying enzymes in foods, whereas heat is. Radiation – sterilized foods are therefore subjected to a mild preheating treatment which is adequate to inactivate undesirable enzymes but does not effect *adversely* the eating quality of food. The severity of the treatment varies from food to food e.g. beef –2 to 5 minutes at 70 °C is adequate.

## Extending Food Irradiation

Under Rule 74 of the PFA Rules (1955), some foods have been permitted for irradiation at different dose levels. However, there is a huge pending list of food items yet to be permitted for irradiation with a dose level up to 10 kGy. The Nuclear Research Laboratory (NRL) of India Agricultural Research Institute (IARI) at Pusa, New Delhi has done work involving various doses of γ-irradiation on button mushrooms, tomatoes, apples, fungi, DNA *penicillium*, pulses and milk. It has also done work on already permitted foods like rice, mangoes, onions and apples.

The man findings in this area have been reported by Roy and Parsed during the silver jubilee of NRL in 1994.

***1.Butter Mushroom.***

Temperature above 15°C were unsuitable for the storage of button mushrooms and a dose of 2.5 kGy was sufficient to preserve it for 9 to 10 days at 15°C as against 4 days for unirradiated buttons (control) under similar conditions.

γ-radiation delays growth and cup opening of the buttons and reduces weight loss as also rottage during storage.

***2. Tomato***

Green mature fruits of three varieties viz. Pusa Ruby, Roma and Pusa Gaurav were studied. All the studied doses (0.5, 1.0, 1.5, 3.0, 3.25 and 4.0 kGy) delayed ripening proportionate to the dose applied. + plus varietal behavior. Fruits of untreated Roma & Pusa Ruby ripen faster than those of Pusa Gaurav.

Abnormal ripening due to irradiation due to irradiation was not observed in the studies. Tomatoes irradiated at 3.0 kGy suffered least due to natural infection during the storage period of 14 days at 23° ± 3°C and 75±5% relative humidity (RH). Studies also indicated that loss in firmness of tomato perceptible immediately after irradiation was dependent on the dose of radiation or on the thickness of the fruit pericarp.

Roma which had a thicker pericarp than the other two varieties suffered comparatively less loss of texture. After 14 days of irradiation, the difference in firmness between treated and untreated fruits was very small at doses of 1-3 kGy. This also suggests that 0.5 kGy is a good dose to delay ripening. For rottage control, however 3.0 kGy is the preferred dose.

On the other hand dose of 3.0 kGy creates loss in ascorbic acid depending on varieties, the reducing and total sugar also were reduced in the fruit juice and Roma and Pusa Gaurav but not in Pusa Ruby – the pH of the juice increased and acidity decreased. However, juice yield was not affected by irradiation at 3.0 kGy.

### *3. Apple*

Decay of apples due to 48 hours old infection *Aspergillus niger, Penicillium expansum, Bortrytis cinera* can be effectively controlled by the combined treatment of γ-irradiation (1.5 kGy), hot water (50°C for 10 minutes) and 1000 ppm aureo fungin or 625 ± 375 ppm benomyl dips.

The organoleptic attributes like colour, flavour, taste or aroma and texture of apples treated with hot water and radiation were dependant on temperature, period of storage and varieties of apples.

The sensory scores of apple given the same treatment as mentioned above and stored for 3 weeks at 10°C also compared favorably with controls. Same loss in taste or texture was observed, after the apples were held at 25°C for 3 weeks.

According to Sanjay Rajput *et. al.* about 20-30% of apple produce is lost within short duration of harvest due to paucity of refrigeration facilities as per demands. He has rightly suggested that by preserving the above lost quantity of apples, India can increase production without growing more apples. Shriram Institute for Industrial Research has observed the effect of γ-radiation on shelf life of different varieties of Indian apples, *viz.* Golden, Royal, Red & Rich – a – Red. Freshly harvested apples were irradiated at doses ranging from 0.1 kGy to 0.75 kGy. The irradiated and unirradiated apples were stored at 4°C, 15°C and at ambient temperature for shelf life study. At various stags of storage at different conditions the apples were evaluated for various physico chemical, microbiological and sensory properties.

Radiation processed apples retained acceptability in terms of various properties as compound to unirradiated apples at 4°C even after six month of storage.

### *4. Milk*

The preservation and changes in quality of cow milk irradiated within two hours of milking with various doses (0.0-2.0 kGy) and stored at temperatures (9±1, 18±2 and 32±3°C) was studied by increase in acidity, clot on boiling (COB), peroxide value as well as organoleptic tests. The results showed that the shelf life of milk is dependent on radiation dose and storage temperature.

The shelf life as judged by COB test, of the samples irradiated at 0.5 and 1.0 kGy was 4-7 hours and 8±1 hours respectively, longer than control when stored at 32±3°C, whereas it was longer than control by 17±1 and 22±3 hours, respectively when stored at 18±2°C

Des Raj and M. K. Roy have found that after irradiation, the milk did not exhibit a positive peroxide value test or the formation of free fatty acid, which are the oxidative and hydrolytic properties of milk fat. However the milk showed increased viscosity, roughness to tongue and palate and little irritating to the gums and lips. This flavor being different from the oxidative and hydrolytic rancidity has been coined as "*radiolytic rancidity.*"

This radiolytic rancidity was found to increase with the increase of radiation dose. The repulsive flavoring components seem to originate from the milk proteins, as after irradiation, the milk became more susceptible to alcohol test than the control, which is the property of destabilised milk protein. Further work is needed to find out whether this radiolytic rancidity can be avoided and irradiated milk is fit for human consumption.

***Source** – DAE – BRNS Symposium on Nuclear Application in Agriculture, Animal Husbandry and Food Preservation New Delhi;* March 16-18, 1994 – p. 154.

According to A.K. Rathour, (Mahaan Proteins Ltd., Mathura) once the economic limitation in the commercial use of the radiation processing is overcome, then preservation by irradiation provides the best option to render dairy products wholesome and safe for long periods and conserve energy. In the wake of world wide consciousness for energy conservation in recent times, radiation preservation method seems to be a viable proposition.

## Inhibition of sprouting of Potatoes, Onions and Carrots

The sterilization process described in earlier pages represents the most difficult application of radiation to food. Inhibition of sprouting presents the other extreme, a process with fewer problems. The factors making it relatively simpler are primarily the two under stated.

1. The required dose is very low and there fore the safety considerations both with respect to operation as well as to product wholesomeness are easily met.
2. Inhibition of sprouting can be achieved by a single application of radiation and does not require subsequent prevention of contamination and special packaging and storage precaution.

The rationale of the process is as follows: -

Immediately after harvest, potatoes are in a dormant state and under ideal environmental conditions, can be stored for many months without sprouting and mould growth under commercial conditions. The ideal environment of about 7°C and 95% relative humidity is difficult to maintain. For this reason potatoes must be treated to avoid sprouting, either by treating with maleic hydrazide or other inhibitors or by irradiation.

Irradiation does produce some damage to the capability of potato tissue to heal quickly after bruises and other injuries as well as the production of some temporary changes to normal respiration of potato tuber.

Nevertheless, these defects fade in significance to other advantages of their irradiation hence are ignored and the treated product is deemed wholesome. In other words, irradiation of potatoes, carrots and onions to achieve sprout inhibition is ready for use permission granted under PFA Act, but economics dictate its suitability

### Premature Maize Ears – safe storage

Maize green ears at milky or early dough stage has better protein quality than mature ones. These are commonly consumed in roasted or boiled form as snacks. During storage, deterioration occurs rapidly in them due to infection by molds and other microorganisms. Besides, sweetness and softness of grains change if consumed after a few days of harvest, thereby reducing consumers preference. Irradiation at a dose of 3.0 kGy for control of molds, particularly *Aspegillus* and *Penicillium* of maize ears has been found to prevent any mold growth at the peduncle and stone of maize even after 45 days of storage. There was no change in constituents viz starch, protein, total sugars.

Source:- H. O. Gupta, M. K. Ray, N. N. Singh, *J. Food. Sci. Technol.*, 2003,40 (5) 512-14.

The detailed data is on pp. 141-143 for insight study.

### Rice

There is an increase in the crystallinity of starch with increments in doses, so also soluble amylase, yellowness in treated grains but showed decreases in water absorption and volume expansion on cooking. The off aroma in irradiated grains becomes perceptible at doses higher than 5 kGy. The changes in colour and aroma persists also on cooking. Upto a dose of 5 kGy the sensory scores of rice, both cooked and uncooked were at or above acceptable limit of score (5.5). The doses of 3.0 and 5.0 kGy are highly effective in reducing fungal population in irradiated grains but in view of

**Applications of Food Irradiation**

| *Type of food* | *Radiation dose in kGy* | *Effect of treatment* |
|---|---|---|
| Fresh foods (fruits and vegetables) | 1.0 (maximum) | Delay maturation, disinfestation |
| Pork carcasses or fresh, non heat processed cuts of pork carcasses | 1.0 (maximum) | Control of *Trichnella spiralis* |
| Dry or dehydrated enzyme preparation | 0.3 to 10 | Control of insects and/or of microorganisms |
| Meat, poultry, fish, shellfish, some vegetables, baked goods, prepared foods | 20 to 71 | Sterilisation. Treated products can be stored at room temperature without spoilage. Treated products are safe for hospital patients who require microbiologically sterile diets. |
| Spices and other seasonings | Upto a maximum of 30 | Reduces number of microorganisms and insects, replaces chemicals used for this purpose. |
| Meat, poultry, fish | 0.1 to 10 | Delays spoilage by reducing the number of micro-organism in the fresh refrigerated product. Kills salmonella, camylobacter and other food poisoning bacteria and renders harmless disease-causing parasites (*e.g.* trichinae). |
| Strawberries and some other fruits | 1 to 5 | Extends shelf life by delaying mold growth |
| Grains, fruits, vegetables and other foods subject to insect infestation | 0.1 to 2 | Kills insects or prevents them from reproducing. Could partially replace post-harvest fumigants used for this purpose. |
| Bananas, avocados, mangoes, papayas, guavas, and certain other non-citrus fruits | 1.0 (maximum) | Delays ripening. |
| Carrots, potatoes, onions, garlic, ginger | 0.05 to 0.15 | Inhibits sprouting. |
| Grains, dehydrated vegetables, other foods | Variours doses | Desirable changes (e.g. reduced rehydration times). |

**Relative Tolerance of Fresh Fruits and Vegetables to Irradiation Below 1 kGy**

| *Dose range* | *Food product* | *Advantages* |
|---|---|---|
| High (irradiation doses upto 0.75 kGy) | Apple, cherry, date, guava, logan, muskmelon, nacturine, papaya, peach, rumbutan, raspberry, strawberry, tamarillo, tomato. | Irradiation doses up to 0.75 kGy were sufficient in controlling spoilage without doing any damage to the fruits composition or taste. |
| Medium (irradiation doses of 0.30 kGy) | Apricot, banana, cherimoya, fig, grapefruit, kumquat, loquat, lychee, orange, passion fruit, pear, pineapple, plum, tangelo, tangerine. | Eliminating harmful bacteria, without affecting overall fruit quality. |
| Low (irradiation doses of 0.15 kGy and 0.25 KGy) | Avocadao, cucumber, grape, green bean, lemon, lime, olive, pepper, sapodilla, soursop, summer squash, leafy vegetables, broccoli, cauliflower. | No effect on compositions. |

**MEAN VALUE OF EARLY DOUGH STAGE RADIATED (R) AND UNRADIATED (UR) MATERIALS AT FOUR STAGES OF STORAGE (% DB)**

| *Storage peroid, day* | *Treatment* | *Weight,g* | | *Starch %* | | *Protein, %* | | *Total sugar, %* | |
|---|---|---|---|---|---|---|---|---|---|
| | | *100 grains* | *100 endosperms* | *Grains* | *Endosperms* | *Grains* | *Endosperms* | *Grains* | *Endosperms* |
| 0 | UR | 17.6 | 13.6 | 68.2 | 70.8 | 14.8 | 14.2 | 2.7 | 2.1 |
| | R | 15.7 | 12.8 | 62.7 | 69.2 | 14.6 | 2.6 | 2.6 | 2.4 |
| 15 | UR | 17.6 | 13.1 | 71.0 | 72.7 | 15.1 | 14.4 | 2.5 | 1.7 |
| | R | 17.1 | 13.7 | 62.0 | 62.5 | 15.3 | 14.9 | 2.9 | 1.8 |
| 30 | UR | 20.7 | 15.1 | 69.3 | 69.6 | 15.0 | 14.6 | 2.8 | 1.9 |
| | R | 16.0 | 12.0 | 61.6 | 69.4 | 14.6 | 13.9 | 2.8 | 1.8 |
| 45 | UR | 18.6 | 14.5 | 64.6 | 66.8 | 16.5 | 14.9 | 3.0 | 1.8 |
| | R | 14.6 | 11.2 | 59.2 | 64.2 | 15.5 | 14.9 | 2.6 | 2.0 |
| | SEm± | 2.2 | 1.5 | 5.6 | 5.9 | 0.7 | 0.4 | 0.3 | 0.4 |
| | CD at 5% | 7.3 | 5.0 | 18.3 | 19.2 | 2.2 | 1.2 | 0.9 | 1.3 |
| | Mean* UR | 18.6 | 14.1 | 68.3 | 70.0 | 15.3 | 14.5 | 2.8 | 2.0 |
| | R | 15.8 | 12.4 | 61.4 | 66.3 | 15.0 | 14.5 | 2.7 | 2.0 |
| | SEm± | 1.2 | 0.8 | 3.1 | 3.3 | 0.4 | 0.2 | 0.1 | 0.2 |
| | CD at 5% | 3.9 | 2.7 | 10.3 | 10.7 | 1.2 | 0.6 | 0.5 | 0.7 |

UR: unradiated; R: Radiated:

changes in colour and cooking qualities 3 kGy is the preferred dose limit of irradiation. The summed up result are tabulated below.

**Colour, Aroma, Texture and crushing pressure of Irradiated Rice.**

| *Irradiation dose kGy* | *Uncooked Rice* colour | *Aroma* | *Texture* | *Av. Crushing* | *Cooked Rice* Colour | *Aroma* | *Texture* |
|---|---|---|---|---|---|---|---|
| 0 | 7.7 | 7.1 | 6.4 | 20.5 | 7.1 | 6.6 | 6.4 |
| 1 | 7.6 | 6.4 | 7.1 | 22.0 | 6.6 | 6.4 | 6.0 |
| 3 | 6.0 | 5.9 | 6.4 | 22.3 | 6.6 | 6.2 | 6.2 |
| 5 | 6.0 | 5.9 | 7.0 | 25.0 | 5.5 | 6.4 | 6.0 |
| CD at 5% | 1.33 | NS | NS | NS | NS | NS | NS |

NS- Not Significant.

Further studies were made to remove or atleast minimise the above organoleptic drawbacks of irradiation. It was found that irradiation at 1 or other lower doses of 0.5 and 0.25 kGy do not effect some of the cooking and eating qualities like length and volume expansion of cooked rice grain Water absorption on cooking of rice and alkali digestibility of uncooked rice are also not affected by irradiation though there is slight reduction in the net consistency.

It was also proved that 0.5 kGy and above leave no surviving insects like *Tribolium* and *Sitophilus* even after 40 days. Consequently a minimum dose of 0.25, maximum of 1.0 and overall average dose of 0.62 kGy have been fixed under rule 74 (4) of PFA Rules is 1955.

Studies showed that a dose of 1 kGy of $\gamma$-radiation kills cent percent of the adult pulse beetle within a week. But the dose of 0.5 kGy required two weeks to achieve the same level of mortality. For a 6 months storage, a dose of 1 kGy was effective for the control of natural infestation of *moong* beans and gram by *Callosorbuchus chinensisalone*. The dose was also sufficient for the management of weevilling in gram and lentil which were additionally infested with *C. chinensis*, rust red flower beetle. *Tribolium castaneum* and lesser grain borer (*Rhizoputha dominicia*). For a short (1.5 months) storage of moong bean and lentil a dose of 0.25 kGy was found adequate. The test of significance for doses of radiation, storage periods and their interactions, however, indicated that 1.0 kGy is the preferred dose for the control of weeviling during storage upto 6 month hence under Rule 74(2)(13) of PFA. Rules 1955, a maximum dose of 1.0 kGy, minimum of 0.25 kGy and an overall average dose of 0.62 kGy has been fixed.

Source: M.K. Roy and H.H. Prasad, *Twenty five years of Nuclear Application in Agricultures Research*, India Agricultural Research Institute, New Delhi, (1994), pp. 208-217.

Ionizing radiations (X and γ-rays) and ethylmethane sulphonate (EMS) have been commonly used for enhancing variation of polygenic traits including flowering and inducing major gene mutation in sayabean[1] and pearlmillet[2] and poppy seed[3].

Considerable progress has been achieved in the utilization of ionizing radiations in the improvement of essential oil crops and several mutant clones with increase in herbage and oil content in *Mentha citrate*[4] and Geranial Jamrosa[5]. Gupta *et al.*[6] investigated mutagenic effect of different doses of EMS and γ-rays alone and in combination in highly inbred live of linalool rich strain of *Ocimum canum Sims*. The $m_1$ generation raised from treated seed did not show any appreciable alteration in plant morphology, oil content and composition. In segregating $M_2$ population, however, a high frequency of both chlorophyll deficient mutations and micromutations (affecting the oil yield and its composition) have been noticed in several treatments especially in the treatment of 10 kR + 0.4%–EMS which exhibited pronounced effect in enhancing the oil content from 0.62% (control) to 0.92% i.e. 50% enhancement. The combined mutagenesis proved to be more effective than individual treatments. Treatment such as 20 kR + 0.8% EMS, 15 kR + 0.6% EMS and 10 kR + 0.4% EMS caused a higher frequency of chlorophyll mutation and micro mutation than the control in that order. This study led to the isolation of several desirable plant types including a line with an increased percent (61±1%) of linolyl acetate.

## REFERENCES

1. Kiang, L.C. and Halloran G.M.; *Mutation Res.,* 1977 43, 223-230.
2. Hanna, W.W. and Burton, G.W. *Crop Sci*, 1985, 25, 79-81.
3. Gille, E.V., Gheorghita, G.I. and Puizary, G.; *Revieu Romain de Biologie vegetable*, 1985, 30, 73–78.
4. Kak, S.N. and Kaul, B.L., *Indian Perfumer*, 1988, 32 (2), 173-180.
5. Singh, C., Paul, S., Lal, S., and Thappa, R.K. *Indian Perfumer*, 1990, 34(3), 190-95.

## Chapter 13

# FOOD IRRADIATION—ECONOMICS

Developing countries situated in the tropical regions have high humidity and temperature. These conditions accelerate the spoilage of food items. Further, these countries generally lack the infrastructure for the better conservation of food items. Chemical treatments of food items are generally harmful hence not appreciated.

Food irradiation has become established as a better alternative for food preservation. It eliminates the use of harmful chemicals for food treatment. Besides, irradiation maintains the original freshness of the foods. Food irradiation is of relevance to India as a developing country to avoid wastage of agriculture of agriculture produce like

(a) sprout inhibition in root crops such as potato, onion, garlic, shallots (small onions) etc.,
(b) delayed ripening in fruits like mangoes, bananas,
(c) disinfection of grains,
(d) reduction of microbial load in spices.

Besides these, sea foods and meat products can increase their shelf-like and reduce microbial load in frozen sea-foods.

Seema Yadav had rightly said that

"above stated all features of radiation applied to food are very satisfactory. One of the main disadvantages is expense. Another might be organization, for it would hardly be possible to set up large numbers of widely scattered plants: both these factors must be balanced against two advantages, the elimination of disease germs both in animals and man, and the saving of vast quantities of food lost by spoilage. More important than expense, however, is safety, and this is why radiation of food is still experimental. All that can be said at the moment (1997), as a result of experimental feeding of animals with irradiated food, is that neither of these hazards production of toxic products/food itself might become radioactive and its consumption dangerous] seem to occur with doses used." She continues:-

"We must, of course, not forget the germs themselves, for they do not readily give up the struggle for existence. They have already shown this by their ability to develop antibiotic resistance..... Man was caught unawares with antibiotic resistance."

Dr. D.R. Bongiwar of BARC had also stated that any process which inflates the price of commodities will be resisted, and if accepted it would be out of sheer compulsion. Therefore, the economics of the process and infrastructure of irradiation technology needs a close scrutiny before adoption as a commercial alternative. Elaboratively, the details of food irradiation facility, its design, infrastructure of process control etc. are its important aspects.

Sources of ionizing radiation have already been detailed. In food processing the following parameters are kept in view before finalization of the design:

(i) products density,
(ii) doses - their range and dose,
(iii) volume of products to be handled,
(iv) radiation source to be used

With these major and several minor variables, it is difficult to evolve an economic and optimum design of a general pürpose food irradiation facility for wide range of treatment processes from sprout inhibition to sterilization in the same facility without compromising the factors like source utilization and uniform dose distribution.

## Economics of Food Irradiation

The economics of food irradiation cost depends on a host of factors specific to each item like

(a) pattern of production,

(b) storage,
(c) distribution,
(d) handling and,
(e) internal consumption or export.

The approximate cost estimates of food irradiation and the actual processing have been worked out by Dr. Bongiwar of BARC and presented during the Golden Jubilee Year of Shriram Institute for Industrial Research Seminar (1999). The details are as under:–

## Irradiation System

It is imperative to choose an irradiation of appropriate size for economical operations. Processor can suffer stiff production cost penalties if too large a plant is run at less than capacity rather than operating on smaller plant at its ideal out put. The costs are fixed according to the size of the irradiator and not on its actual product handling volume. In other words, a large irradiator treating small volumes of products has less out put over which spread its high fixed costs.

## Radiation Power Cost

The initial cost of radiation power can be calculated on the basis of the quantity of radiation source required for treatment purpose at the time of installation and can be calculated considering the following parameters:-

| | |
|---|---|
| 1. Treatment dose | D in kGy |
| 2. Processing rate | T in tonnes/hour |
| 3. The use fraction | E of the source energy effectively absorbed in the product in the irradiation system. |

These three parameters may be related to Q, the radiation power required in Kilowatt (Kw) and is given by the equation

$$Q = TD/E \text{ (Kw)} \qquad \text{(i)}$$

The value of E under practical conditions, usually lies between 0.05 to 0.4. Such a low value is attributable to the high penetrating nature of the γ-radiation and isotopic nature of emission of radiation energy from isotopic sources.

Also a considerable fraction of the radiation is wastefully absorbed in the packaging materials, structural components and conveyor materials and also leaks out through space between packages or between the loosely packed food products themselves. The above equation (i) above can also be rewritten to give the required source power in terms of curies of $^{60}$Co activity (1 Kw = 67.8 kilo curies of $^{60}$Co). The required activity A is related by the following equation:

A (kilo curies of $^{60}$Co = 5.3

(T.kg/hr) x D (kGy)/E (ii)

A (megacuries) of $^{60}$Co = 5.3

(T.Ton/hr) × D(kGy)/E (iii)

The energy requirement or the treatment dose in terms of kGy for various processes of interest in food preservation.

kGy = 1 kjoule/kg = 240 cal/kg of radiation energy. Given the price of $^{60}$Co, one can also calculate the basic energy cost as shown below.

**Energy Requirement/Treatment Dose/Cost of Power***

| *Process* | *Range of dose(kGy)* | *Average Dose (D) (kGy)* | *Approx. Cost of Radiation Power* |
|---|---|---|---|
| Sprout inhibition (root crops) | 0.06-0.15 | 0.10 | .0.1P/kg |
| Disinfection (grains) | 0.30-0.60 | 0.45 | 0.5 -do- |
| Pasteurization/Shelf Life Extension (fish) | 3.0-6.0 | 4.5 | 5.0 -do- |
| Microbial Reduction (spices etc.) | 5.0-10.0 | 7.5 | 25.0 -do- |

Unit cost of radiation power at Rs. 36/kwh or $^{60}$Co cost of Rs. 15/ curie installed

P = price

* On the assumption of full utilization of radiation power E = 1

*Source*:- K. Krishnamurthy & D. R. Bongiwar, *Indian Food Industry*, 1987, 6, 6-7.

In gamma irradiation facilities major recurring costs (R) are due to the need for periodic addition of sources to compensate for the continuous decay of radiation sources, operating, maintenance and consumables. The maintenance costs in gamma facilities are usually minimum. In $^{60}Co$ irradiators about 12% of the initial activity may be required to be added annually to compensate for the decay. In addition to the costs of $^{60}Co$ the cost of transport of these sources in heavy lead containers to the site installation of $^{60}Co$ in the facility is also substantial. The cost of labor, however, may depend on the capacity utilization of the plant and the seasonal nature of the products treated.

It is necessary to select an appropriate size irradiator for economic operation. Processors can suffer stiff production cost penalties if too large a plant is run at less than capacity rather than operating a smaller plant at its ideal through put. A large irradiation treating small volumes of products has less output over which to spread its high fixed costs.

An extensive analysis of available information on the cost of shield and conveyer systems used in a variety of irradiators both for demonstration and commercial radiation processing and our own experience with the design, installation and operation of irradiation facilities have indicted the following simplified approach:-

## Cost Estimates

As hinted above, these depend upon the following two basic data:

(A) The installed $^{60}Co$ activity (A) in mega curies.

(B) Maximum designed capacity of the irradiator (Am) expressed in mega curies.

I Capital Cost Estimate (I)

Total investment (I) = Cost of source (S) + Cost of building and plant (B)

$$S = Aip, \text{ where}$$

Ai is the strength of $^{60}Co$/curie source initially installed in terms of numbers of curies in millions;

P is the price of $^{60}Co$/curie inclusion of transport and installation cost at site.

$$B = 7.5\ Am^{0.3}, \text{ where}$$

Am is the maximum design capacity of the radiation plant rated in terms of number of curies in million at the highest through put requirements (applicable in the range 0.1 to 5 millions curies of $^{60}Co$.

As stated above

$$I = S + B \text{ or } = Aip + 7.5Am^{0.3}$$

II Running Costs (R)

R comprises of fixed (F) as well as variable costs (V), therefore R = F + V, where

F in million Rs = cost of source replacement + depreciation - as well as maintenance + tax etc.

This may be approximated by the expression

$$F = 0.29\ Aip + 1.08\ Am^{0.3} \text{ and}$$

$$V = \text{Labor cost + overheads.}$$

This may be again approximated by

$$V = B\ 6200\ A \times Ns \times Nm \times 10^{-6} \text{ (million), where}$$

$$A = Aip,$$

Ns = no. of shifts operated daily and

Nm = no. of months/year the plant is operated.

This formula by Bongiwar provides estimates (+50% ····· 30%), valid for the range of activity from 0.1 to 5.0 mega curies of $^{60}Co$ installed. However, the value of $B = 7.5Am^{0.3}$ was based on prices prevalent in the county 1984-1985. Annual increase of about 10-

15% in the cost of labour and material may be considered for future estimates.

On the above basis, Bhabha Atomic Research Center (BARC) has constructed a demonstration facility "PROTON Irradiator" for potatoes and onions at Katamgaon/Lasalgaon in Niphod Taluka of Nashik Distinct where huge quantities of onions are stored for marketing in Maharashtra. Though the exact cost is not available, however, Dr. Bongiwar had estimated the facility cost around Rs. 70 millions in 1999. This facility shall save onions worth Rs. 50 million/annum which is about 20% of the total production which used to go waste due to sprouting.

Broad of Radiation & Isotope Technology (BRIT), a constituent unit of the Dept. of Atomic Energy (DAE) established to promote the applications of radiation and isotope technology, has set up a commercial demonstration plant for radiation processing of spices at Vashi, Navi Mumbai close to the wholesale spice market.

It started with about 3000 tonnes of (variety of) spices a year and has probably reached the targeted capacity of 12000 tonnes. It has arrangements to augmentation of $^{60}$Co depending upon the market demand. It functions round the clock fail-safe automatic operation. Tote box size is (59 cm–l) x (45cm–w) x (110cm–h)). About 30 tonnes of spices (whole and ground) can be processed/ Day at maximum $^{60}$Co loading of one million curies.

The irradiation facility at Nashik can also be usable for other foods products like ginger, garlic, shallots, wheat products like *atta*, *maida*, *sooji*, basmati rice, dry fruits like raisins, figs, and dried dates, meat and meat products including chicken. This wide application helps the facility to be operational for 100% plant utilization thereby enhancing the profits of radiation processing.

The technical and economic factors which favor $^{60}$Co $\gamma$-rays are as under:

*(1) Product Box Sizes:* These sizes determine the dimensions of the tote or carrier. Ideally the product boxes fit into the tote or carrier, as the case may be, such that 100 per cent of the us-

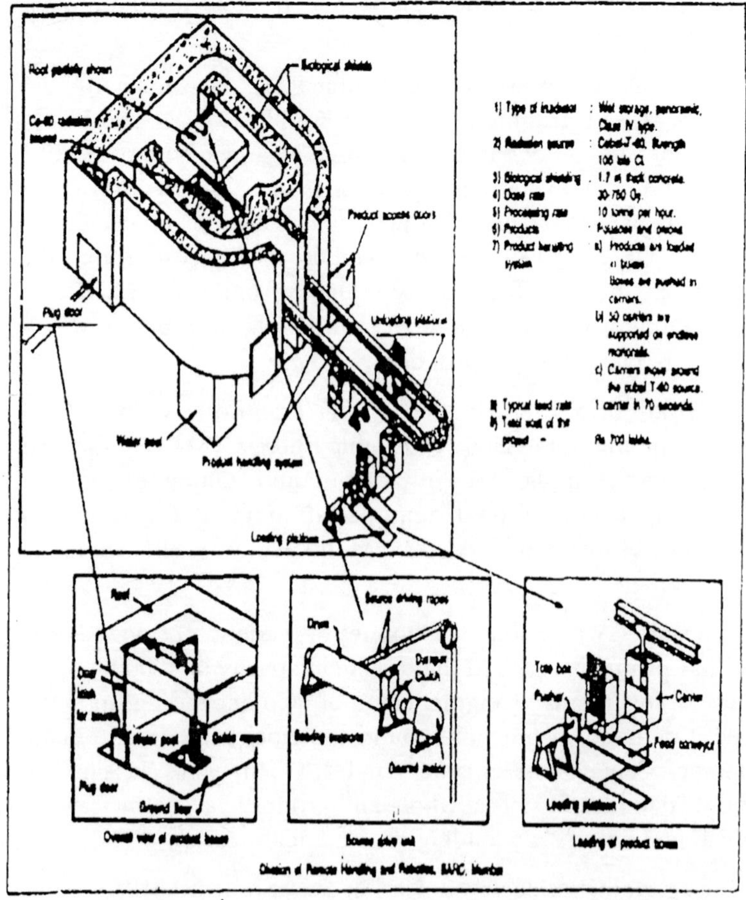


**Fig. 12.1: Proposed Proton Irradiation (potatoes and onions)**

able volume is utilized. The design engineer strives for this ideal otherwise any γ-radiation in volumetric loading efficiency directly results in an increase in the amount of cobalt required.

(2) *Product Densities:* The density of each product plays an important role in irradiation processing. Because of the extremely high penetration of the γ-rays from $^{-60}$Co and the irradiation of the product from several different angles as it passes through the source pass mechanism, the shadowing effect of small high

density metallic parts may be in the product box is negligible. The throughput of an irradiator is a function of product density. Due to attenuation, high density products have a lower volume throughput. Whenever products of different densities are processed together, the dwell time of the product in the irradiator must be set according to the highest density product in the irradiator must be set according to the highest density product to ensure that all products absorb the minimum absorbed dose. When changing from the one density grouping to the other, there may be some cobalt utilization but if there is sufficient product in each density grouping, improvement of Co utilization resulting from the higher throughput of lower density products will more than off set the loss.

(3). *Product Thorough put:* To maximise the use of $^{60}$Co sources, the irradiator be operated around the clock, 7 days/week. Because of routine maintenance, source replenishment etc., it is impossible to attain 100 per cent utilization efficiency, but 8000 hours ( 91% of annual hours available, 91.324% to be more precise) should be readily achievable.

(4) *Minimum Sterialization Dose:* Until recently, the standard minimum dose was 2.5 marads, except in Scandanavia where it is still 3.2 marads. If the minimum dose can be lowered through improved manufacturing practices, a direct improved utilization of $^{60}$Co and corresponding saving will result and shall lead to lower cost of irradiation.

(5) *Maximum Dose and Dose Uniformity Ratio:* Typically dose uniformity ratios between 1.2 and 1.5 can be expected for irradiation designed to process medical disposable. But in food irradiation plant, the overdose ration is generally higher (even upto 2.0) so that higher throughputs are achieved.

Partial Listing of Some Countries and Foods Approved For Irradiation

| *Country* | *Food Products* |
|---|---|
| Argentina | Potatoes, strawberries, onions, garlic |
| Belgium | Potatoes, strawberries, onions, garlic, shallots, paprika, pepper, gum arabic, 78 spices |

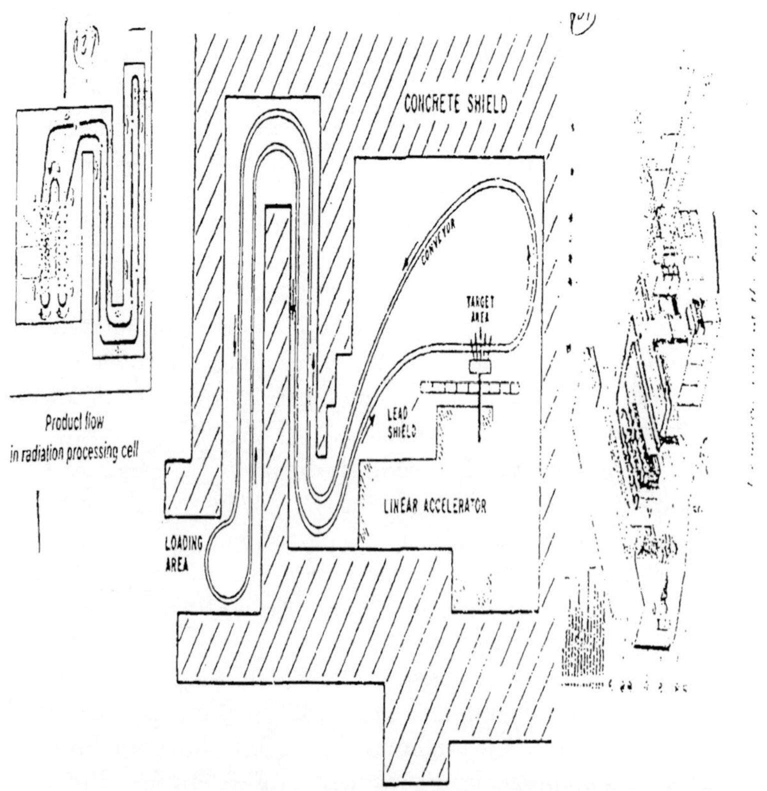


**Schematic representation of the path taken by foods irradiation in the United States Army Natick Linear Accelerator Facility (Based on blueprints provided by United States Army Natick Labs).**

| *Country* | *Food Products* |
|---|---|
| Bulgaria | Potatoes, onions, wheat floor, poultry, cod and haddock, spices and certain dried vegetables |
| Canada | Potatoes, Onions, wheat floor, poultry, cod and haddock, spices and certain dried vegetables |
| Finland | Spices, herbs, hospital meals |
| Chile | Potatoes, papaya, wheat, chicken, onions, rice, fish products, spices |

| *Country* | *Food Products* |
|---|---|
| France | Potatoes, onions, garlic, shallots, spices, dried fruits and vegetables |
| Germany | Hospital Meals |
| Israel | Potatoes, onions, poultry, 36 spices, fresh fruit and vegetables |
| India | Onions, potatoes, frozen seafoods spices, rice semolina, (sooji or rawa), wheat atta, maida, mango, raisins, figs, dried dates, ginger, garlic, shallots, meat and meat products including chicken. |
| Czechoslovakia | Potatoes, onions, mushrooms |
| The Netherlands | Asparagus, cocoa beans, strawberries, mushrooms, hospital meals, potatoes, shrimp, onions, poultry, soup green, fish fillets, frozen frog legs, rice and ground rice products, rye bread, spices, endive, powdered batter mix |
| Philippines | Potatoes, onions, garlic |
| South Africa | Potatoes, onions, garlic, chicken, mangoes, strawberries, dried bananas, avocados, beans |
| Spain | Potatoes, onions |
| Thailand | Potatoes, onions, garlic, dates, wheat, rice, fish, chicken |
| U.S.S.R | Potatoes, grain, fresh and dried fruits and vegetables, dry food concentrates, poultry, onions, prepared meal products |
| U.K. | Hospital meals |
| U.S.A. | Wheat and wheat flour, potatoes, spices, pork, fresh fruits and vegetables |
| Yugoslavia | Cereals, potatoes, garlic, poultry, dried fruits and vegetables |

Factors Affecting the Sterilization Process

| *Factor* | *dry heat* | *Moist heat* | *Formaldehyde* | *Ethylene oxide (ETO)* | *Radiation* |
|---|---|---|---|---|---|
| Process parameters | Temperature time, pressure | Temperature, pressure, time, vacuum | Temperature, pressure, time, vacuum, humidity | Temperature, pressure, time, vacuum, humidity, ETO concentration | Time |
| Retention of sterilient | No | Yes | Yes | Yes | No |
| Post sterilization | No | Yes | Aeration to remove absorbed formaldehyde | Aeration to remove absorbed ETO | Nil |
| Residual toxicity | Nil | Nil | Yes, due to HCHO | Yes due to ETO, ETG, ETCH | Nil |
| Interaction with product | Oxidation, degradation | Hydrolysis | Polymerization | Hydroxyl ethylation | Radiolysis |
| Sterility test | Required | Required | Required | Required | Not necessary |
| Quarantine | Nil | Nil | 14 days(minimum) | 14 days(minimum) | Nil |
| Product density | Affects | Affects | Affects | Affects | Affects to some extent |
| Penetration problems | Exist | Exist | Very Poor | Exist | No problem |
| Choice of packing | Narrow | Narrow | Narrow | Narrow | Wide |
| Sterilization in final packaging for shipping | Not possible | Not possible | Not possible | Not possible | Possible |
| Type of Process | Batch | Batch | Batch | Batch | Continuous |
| Environmental pollution | Nil | Nil | Yes | Yes | Nil |
| Reliability | Fair to good | Good | Fair | Fair to good | Excellent |

## Why Choose $^{60}$Cobalt Gamma Radiation for Sterilisation

| *Consideration* | *Gamma Radiation* | *Ethylene Oxide* | *Steam* | *Electron Beam* |
|---|---|---|---|---|
| Product design | No restriction | Limited sealed cavities | No sealed cavities | Low density product |
| Material of construction | Some plastics discolor, some fibers degrade | Most materials OK | Some plastics melt | Some plastics discolor, some fibers degrade |
| Product packaging | No restriction, no stress, on seals, nonpermeable, OK | Permeable materials seals must withstand pressure, vacuum | Permeable materials expansion of package | Thin product packaging, no restriction, no stress on seals nonpermeable, OK |
| Parameters needing control during process | Time | ETO concentration, vacuum, pressure, temperature, relative, humidity, time | Vacuum, pressure, temperature, relative humidity, time | Time |
| Reliability of process | Excellent | Fair | Excellent | Excellent |
| Poststerilisation | Dosimetric release | Biological indicator | Parametric release | Dosimetric release |
| Quarantine period | No quarantine required | 3-14 days | 0-7 days | No quarantine required |
| Poststerilisation | None required | Aerate to remove residues | Dry product | None required |
| Economics | Good on low and high volumes, high capital investment | Good on low and high volumes, high capital investment | Good on low and high volumes, inexpensive operations | Good on low and high volumes, high capital investment |

## Technical Advantages of Food Irradiation

### *Techo-economic Advantages*

(a) *An alternative to fumigation of food* - Fumigation of food ingredients with various chemicals like ethylene bromide (EDB), methyl bromide (MB), ethylene oxide (EO) is either prohibited or is being increasingly restricted for health, environmental or occupational safety reasons. Purpose was insect disinfection. Both these bromides cause depletion of the ozone layer. Under the Montreal Protocol (an international treaty for the regulation of ozone depleting substances worldwide and under the auspices of UN Environmental Programme) chemical substances which cause depletion of the ozone layer had to be phased out by the 31/12/2000. The maximum permissible limit for the ozone depleting potential is 0.2, MB had 0.7.

Thus the food industry will be deprived of these broad spectrum fumigants used for controlling insect infestation of food.

Irradiation has been demonstrated as an effective alternative to the fumigants mentioned above because it does not leave any residues in the products. It is an alternative to fumigation with EDB/MB of fresh fruits and vegetables to overcome quarantine barriers. According to IAEA (1991b), 0.15 kGy dose is enough as quarantine treatment of fresh horticultural produce against fruit-fly of the *Tephritidae* found and 0.3 kGy against other insects species.

Commercially also, irradiation has increasingly replaced fumigants to ensure hygienic quality of spices and vegetable seasonings because it has broad spectrum and is more competitive in cost than other processes.

(b) *Most Suitable for Solid Foods* : For liquid foods thermal pasteurization is a widely accepted means of terminal treatment but not for solid foods like meat, poultry, sea food and dehydrated ingredients or fresh foods requiring maintenance of their raw characteristics for ready market acceptability. Sanitizing such foods chemically with attending toxic residues is not acceptable as stated at (a) above.

It is unescapable, rather obligatory that raw foods of animal origin, particularly poultry and pork must be free from pathogenic microorganisms and parasites, particularly *Salmonella, Campylo bacter, Trichinella, Toxoplasma* etc. *Vibrio* spp., especially *V. cholerae, V. parahaemolyticus* and *V. vulnificus* also have caused several disease outbreaks and deaths in recent years due to consumption of contaminated water and food. Eating raw fish especially oysters and clams was the mode of transmission of this disease. *V. parahaemolyticus* which is ubiquitous in the marine environment is an important cause of food-borne infection from the consumption of raw or improperly cooked seafood in several countries. *Vibrio spp.* have little resistance to irradiation hence 1 kGy dose inactivates all these. Other pathogens are *Escherichia coli* 0157: H7, *Listeria monocytogenes.* The later is present in several ready-to-eat foods like cheese, processed meats, pates, solid spreads etc. Some strains of *L. monocytogenes* have developed resistance to antibiotics such as tetracycline, erythromycin, contrimoxazole and clindamycin, *L. monocytogenes* can resist heat, salt, nitrate and acidity much better than many other pathogenic micro organisms as well as being able to multiply slowly at temperature as low as 1 °C.

A number of food-borne parasitic diseases such as trichinosis, toxoplasmosis, taeniasis, opithorchiasis, etc., which are prevalent in a number of developing countries, can be controlled by irradiation of food thus making these parasites non-infective without causing significant changes in the physico-chemical or sensory properties of the treated food.

*(c) An energy Saving Process* : The energy required for canning, refrigeration and/or frozen storage is quite large in proportion to irradiation to get the same results. For example, the total energy used for refrigerated raw, cut up chicken is 17760 kJ/kg; for frozen, raw cut up chicken including 4 ± 1 weeks of frozen storage is 46600 kJ/kg. In comparison, refrigerated and irradiated, raw, cut up chicken requires a total energy of 17860 kJ/kg. Thus irradiation together with refrigeration could be used as a substitute for frozen storage for distribution of raw food, e.g. chicken, where feasible like in domestic markets with a

substantial saving in energy and cost. CFC refrigerants which are widely used by the refrigeration and food industries will not be available for use due to Montreal Protocol beyond 2000 AD resulting in higher cost of refrigeration of food. This is unfavorable in India and other developing countries. The cold chain (from harvesting, processing, distribution and refrigeration storage at home) simply does not exist and is too costly for the majority of the population. Alternatively irradiation may reduce dependence on refrigeration of food for the following applications.

(i) *Replacing frozen distribution of food by irradiation in combination with chilling* : A Brazilian large poultry processor has shown that using this temperature saves $ 0.18/kg of poultry with sufficient shelf life for marketing.

(ii) *High quality, shelf-stable food of animal origin* : Meat, sea food, sausages etc., often marketed under frozen conditions need tremendous amount of energy. When sterilized by combined treatment of blanching to inactivate proteolytic enzymes, freezing and irradiation with doses 40 ± 10 kGy renders them equally shelf-stable, needs only modest energy requirements. Irradiated beef steak, corned beef, turkey slices, ham, pork and sausages have been used in many manned space flights. Radiation sterilized cooked salami, shrimp, codfish, cake, bacon pork chop, meat loaf, smoked Viennas, roast chicken, country sausages, beef curry, roast beef etc. have been perfectly developed.

(d) *Cost Effectiveness* : Irradiation reduces food-borne diseases, post harvest losses, making food trade easier. The increasing incidence of food-borne diseases have not only affected health but reduced the economic productivity of population in both advanced and developing countries (WHO, 1984). In USA alone, economic losses from *trichinosis, toxoplasmosis, salmonellosis, carpylobacterious and beef tapeworm in 1985* was estimated to be more than US $ 1.5 million. *Salmonellosis* alone costs $ 2,540 million annually, *toxoplasmosis* upto $ 4,400 million a year. In Thailand the annual cost of human infection with liver fluke *Opisthorchis sinensis* is estimated at about $ 99 million. Low dose irradiation (4 ± 1 kGy) is effective in inactivating non-spore forming pathogenic bacteria and 1 kGy for food-borne parasites mentioned above without causing significant change in physico-chemical and sensory properties of the product.

Irradiation can replace imported fumigants to combat insect infestation economically in cocoa beans, sprouting in potatoes, onions, yams, and garlic can be stopped; mushrooms and asparagus can be curtailed by irradiation. It is not costly also. In Pakistan, the cost of irradiating potatoes and onions is estimates to $4-5/tonne with a net benefit of 40%. In Syria such irradiation yields a net benefit.

In India, oil seeds are the second major agricultural crop next to food grains. Certain pre-harvest events or factors cause drastic post harvest losses. Oil seeds crops, damaged by pest (especially by insects and diseases) give lower yield of oil and protein than healthy uninfected seeds. Consequently the value of the extracted oil is lowered both in terms of quantity and quality.

High energy $\gamma$-radiation has been used for disinfection of oilseeds, groundnut seeds in particular.

Groundnut (GN) seeds were irradiated to a dose of 0.25 kGy. The extracted oil from the above irradiated (GN) seeds had no change in the processing parameters. Chemical composition of the oil remained unchanged. Amit *et. al.* found this a fit case for commercialization of the irradiation of oil seeds, so that their shelf life increases and insect infestation is arrested.

Amit P Pratap *et. al.*, *J. Oil Tech. Asson. India.* 2003, 35(4), 151-159.

Chapter 14

# PACKAGING OF IRRADIATED FOODS

## Introduction

Packaging plays an important role in facilitating the irradiation processing, in protecting irradiated foods from recontamination and in maintaining the hygienic and compositional quality of the foods. Radiation sterilization destroys micro-organism and thus prevents their migration from the packaging to its contents, where by the later should remain completely unaffected by the radiation process. In other wends, no unwanted changes should take place in the contents themselves.

The packaging may consist of a variety of materials e.g. metal, metallised plastic, plastic or glass. Radiation sterilization offers the advantages that the inside of closed packaging also gets sterilized, this being one of the prerequisites for the aseptic packaging of products, or for the packaging of products which may not be contaminated by extraneous organisms.

It necessitates to consider the influence of irradiation of packaging materials. The essential characteristics of packaging being effective then the irradiation should neither compromise the functional properties of the packaging materials non-facilitate the migration test to be positive i.e. radiation stability is the key consideration in the selection of packaging materials.

Packaged food is comprised of different types of macromolecules. The goal of irradiation is to maximise damage to the DNA of the contaminating bacteria and to minimise damage to the structural polymers of the packaging. This paradoxical goal can be accomplished if at least one of the following two conditions is met.

(i) The definition of damage threshold for the two type macro-molecules is substantially different;
(ii) The radiation durability of the two types of macromolecules is substantially different.

The interaction of ionizing radiation with matter takes place via transfer of energy to the electrons in atomic or molecular orbitals, resulting in their displacement. This displacement can eventually result in bond scission, which is the main concern with respect to the damage to polymers. Since most commonly used polymers comprise primarily of carbon, hydrogen, nitrogen and oxygen atoms and have molecular orbits of similar size, their durability to radiation can be classified in a simplified way according to the nature of there orbital. These molecular orbitals, associated with the polymer backbone, play the major roles in the resistance of polymers to scission. The radiation durability of these orbitals serves as a general ranking of the radiation durability of polymer families.

## DNA Damage Mechanism with Food Irradiation

When microbes present in the food are irradiated the energy from the rays is transferred to the water and other molecules in the microbe. The energy creates transient chemicals that damage the DNA in the microbes, causing defects in the genetic instructions. Unless it can repair this damage, the microbe will die when it grows and tries to duplicate itself. Disease causing organisms differ in their sensitivity to irradiation, depending on the size of their DNA, the rate at which they can repair damaged DNA, and other factors. It matters if the food is frozen or fresh, as it takes a higher dose to kill microbes in frozen foods. The size of DNA "target" in the organism is a major factor. Parasites and insect pests, which have large amount of DNA, are rapidly killed by extremely low doses of irradiation, D-values of 0.1 kGy or less. It takes more irradiation to kill bacteria, because they have a somewhat smaller DNA, with D-values in the range of 0.3 to 0.7 kGy. Some bacteria can form dense hardy spores, which means they enter a compact and inert hibernation state. It takes more irradiation to kill a bacteria spore, with D-values on the order of 2.8 kGy. Viruses are the smallest pathogens that have nucleic acid and they are in general resistant to irradiation to doses approved for foods. The prion particles associated with bovine spongiform encephalopathy (BSE, also known as mad cow disease) do not have nucleic acid at all, so they are not inactivated by irradiation, except at extremely high doses. This means that irradiation will work very well to eliminate parasites and bacteria from food, but will not work to eliminate viruses or prisons from food.

***Packaging Materials and Irradiation:*** During processing, packaging materials are exposed to radiation so stability is a key consideration in material selection. In radiation processing of food, the product often has to be pre-packaged to prevent microbial recontamination during and after irradiation. The JECFA (Joint Expert Committee on Food & Agriculture) published a report way back in 1981, and therein concluded the irradiated food upto a dose level 10 kGy (kiloGray) was safe and nutritious. In 1983 the Codex recommended standards for irradiation of food as discussed earlier.

The growing use of food irradiation has created a need for suitable approved packaging materials which can withstand radiation processing. It is important to select materials in which chemicals formed as a result of radiation treatment do not migrate and interact with the food, affecting its organoleptic and toxicological aspects. It is also important to select materials in which the physical properties are not altered to the extent that they cannot resist damage during commercial production, shipment and storage. Radiation treatment of food may be classified broadly into two categories:

(i) Processes requiring dose less than 10 kGy such as extending the refrigerated shelf life of meats, inhibiting sprouting of onion, potatoes, insect disinfection and many others:

(ii) Processes requiring dose from 25 to 40 kGy for production of commercial sterility.

In radiation processing of food, $\gamma$-radiation is most widely used because of its high penetration power. Electron beam irradiation can also be used for certain specified foods and packaging materials.

The two major effects of ionizing radiation on polymers are cross linking and scission of the polymer chains. In general, the two competing effects occur simultaneously, and the predominating effect depends upon the structure of the polymer. The net effect of cross linking reactions is to modify the physical properties of the material such as :

(i) increasing the tensile strength,
(ii) hardening,
(iii) changing the solvent resistance, and
(iv) decreasing the impact strength.

Chain scission involves rupturing the molecular bonds, thus leading to the :

(i) evolution of gases and
(ii) change in extractables.

For the most part, the dose at which radiation effects become significant for the less radiation resistant materials-polypropylene, polyvinyl chloride, cellulose, poly vinylidene chloride, is 10 kGy.

Where single films have been approved for irradiation, frequently they do not meet the packaging needs. For example, the packaging films used for red meats usually have both high moisture and high oxygen barriers. In such cases better results are achieved with multilayer rather than with single films. For this very reason the industry should seek approval for suitable packaging materials for the radiation processing of foods. For guidance, thickness of pouch/packaging materials required for food irradiation have been tabulated here, so also radiation stability of some materials used in food packaging.

**Thickness of Pouch/Packaging Materials Required for Food Irradiation**

*(Components of multilayer flexible pouches used successfully for sterilizing food with ionizing energy and for protecting the food from subsequent contamination)*

| *Pouch No.* | *Pouch film components* | *Film component thickness, microns** |
|---|---|---|
| 1. | Polyethylene terephthalate | 13 |
| | Aluminum foil | 9 |
| | Polyethylene terephthalate | 13 |
| | Polyethylene, 0.96 gram per milliliter | 80 |
| 2. | Polyethylene terephthalate | 13 |
| | Aluminum foil | 9 |
| | Ethylene-butene-1 copolymer | |
| | Polyisobutylone blend (70-30) | 80 |
| 3. | Polylminocaproyl | 25 |
| | Aluminum foil | 25 |
| | Ethylene-butene-1 copolymer, | |
| | 0.950 gram per milliliter | 80 |
| 4. | Polyethylene terephthalate | 13 |
| | Aluminum foil | 13 |
| | Polypropylene-ethylene | |
| | Vinyl acetate copolymer (94-6) | 80 |
| 5. | Polyiminocaproyl | 25 |
| | Aluminum foil | 9 |
| | Polyethylene terephthalate | 13 |
| | Polypropylene | 80 |

* The 6-micron adhesive between any two layers is epoxy-polyester with the reaction product of trimethylolpropane and 2, 4-toluene di-isocyanate.

**Radiation Stability of some materials used in Food Packaging**

| *Material* | *Relative Stability* | *Predominant reactions (In absence of oxygen)* | *Observed effects* |
|---|---|---|---|
| Cellulose plain and coated Papers, Kraft paperboard, Cardboard, jute etc. Coated-cellophanes Vegetable parchment Cellulose acetate | * | Chain scission | Yellowing, loss in strength, eventual disintegration, embrittlement and reduction in elongation at break in cellulose derivatives. |
| Poly (vinylidene chloride) (PVDC) | * | Chain scission | Darkening, odour, HCl evolved. |
| Poly (vinyl chloride) (PVC) | ** | Degradation | |
| Vinyl chloride-vinylidene Chloride copolymers (Saran) | ** | Crosslinking scission | Yellowing-brown on standing, HCl evolved, |
| Vinyl chloride-vinyl acetate copolymers | ** | unsaturation | reduction in tensile strength, decreased elastic modulus. |
| Polypropylene, Propylene/ethylenevinyl Acetate copolymer | * | Crosslinking scission, gas evolution | Embrittlement, yellowing. |
| Ethylene-vinyl acetate Copolymer, Ethylene-butene-1 copolymer | ** | Crosslinking | Reduced elongation at break |
| Polyethylene | *** | Crosslinking | Discoloration, hardening, slightly increased gas evolution, tensile strength, resistance to creep and stress crackling, negligible change in permeability, flexibility and tear resistance. |
| Polycarbonate | *** | Chain scission | Embrittlement at high doses. |
| Nylon-6 | *** | Crosslinking | Hardening, decreased solubility, notch impact strength, elongation at break, increased heat resistance and modulus. |

The net effect of cross-linking reactions is to modify the material for properties like.

(a) tensile strength enhancement,
(b) changing the solvent resistant,
(c) decreasing the impact strength and
(d) hardening

The conditions under which food products are irradiated bear a significant influence on the behavior of packaging material. For instance, volatile compounds evolved when a low density polyethylene is irradiated under vacuum consist mainly of hydrogen + saturated and unsaturated hydrocarbons.

On the other hand when irradiation is done in the presence of oxygen the volatiles evolved are aldehydes, ketones, carboxylic acids and saturated hydrocarbons.

This necessitates the determination of chemical and physical (with respect to radiation dose, temperature and atmosphere) of the packaging material under the expected condition of use.

Modern packaging needs involve complex structures using two or more films with different barrier properties. Barrier to moisture, gases, aroma or flavour, light are the necessities for certain packaging. Aroma and flavour – barrier materials are needed to prevent extraneous odors from contaminating the product. Some foods are spiced or flavoured artificially and the packaging material should be able to retain this flavour.

The most commonly used aroma- flavour barriers are nylon and ethylene vinyl alcohol.

Moisture barriers are essential in most packaging applications to keep the desired level of moisture in the product. Propylene, high-density polyethylene (HDP) and poly- vinyl chloride (PVC) are good barriers.

Nylon or polyester are used as modern gas barriers. High barrier structures contain either PVC or polyvinyl alcohol.

Besides PVC, polyethylene and polypropylene (PP) contain antioxidants like phenolic and arylphosphate compounds.

Organotin compounds are added to PVC to protect against thermal degradation. Studies have shown than some of these additives get degraded by irradiation and the degradation products have a great propensity to migrate into food. It has also been proved that in some polymers loss up to 30% of hindered phenol antioxidants occur on exposure to a radiation dose of 30 kGy. It is, therefore imperative that during the manufacture of such types to select the additives which are stable to radiation lest any migration may occur to the contained food.

It is heartening that new antioxidants and nucleating agents are available now which not only are stable to radiation but also change the morphology of the polymer in which they are in corporated. These have lessened the effects of PP and PVC discoloration earlier caused by radiation to a great extent. Effect on the mechanical properties of PP compositions due to incorporation of antioxidant in combination with nucleating agent at various sterilization doses are tabulated below.

**Mechanical Properties of Modified PP vs Virgin PP at different Irradiation Doses.**

| | *Mechanical Properties* | *Unexposed* | *Exposed* 25 kGy | 50kGy | 75kGy |
|---|---|---|---|---|---|
| Virgin PP | Impact strength (joule/meter) | 19.32 | 12.43 | 8.29 | 5.39 |
| | Tensile Stress (kg/cm$^2$) | 341.7 | 330.9 | 260 | 170 |
| | Elongation(%) | 16.1 | 9.80 | 5.21 | 3.9 |
| PP with Additives | Impact strength (joule/meter) | 20.31 | 19.93 | 14.2 | 9.2 |
| | Tensile stress (kg/ cm$^2$) | 374.3 | 368.0 | 310.0 | 230.0 |
| | Elongation (%) | 16.0 | 15.5 | 12.2 | 8.2 |

PVC accounts for over 30% of the plastics used for food packaging and other food contact applications. It is used as flexible films containing plasticisers to package food items such as fresh meat, fish, fruits and vegetables or in the rigid, unplasticized from as containers and bottles. PVC seals are also widely used in gaskets and stoppers for sealing jars and bottles. In aseptic packaging there is a great need of materials capable of withstanding the required radiation doses.

Government of India, vide its Gazette notification of March 2, 1991 (Part 11 - Sec 3), has recommended following polymeric packaging films in radiation processing of foods upto 10 kGy.

(i) Nitrocellulose or vinylidene coated cellophane.
(ii) Wax coated paperboard.
(iii) Glassine paper
(iv) Polystyrene.
(v) Rubber hydrochloride.
(vi) Vinylidene / vinyl chlorides.
(vii) Polyethylene.
(viii) Polyethylene terepthalate
(ix) Nylon 6 and 66
(x) Vinyl - chloride / acetate
(xi) Vegetable parchment

It has also been recommended that flexible packaging materials like flexible plastic containers are highly suitable for packaging of irradiated-foods by virtue of their low density. Films over 76 micrometer thick are satisfactory for preservation of irradiated foods.

From the above, it is obvious that plastic packaging is vital for food sterilized by ionizing radiations.

Some times all the desired properties are not inherent in a single polymer packaging materials. By co-extruding different materials one gets the materials of desired requirements. The materials already cleared do not fulfill all the needs of modern food packaging hence search for new material is never ending.

## Conclusions

On the basis of recent advances it has been concluded that:

(i) The currently used trilaminate pouch developed by Natick with polyethylene as the food contacting layer (which is approved) is of proven safety, based on experience and long-term wholesomeness testing.

(ii) The concept of double packaging, which provides a single approved layer in contact with the food, over wrapped with a laminated package with the requisite physical properties, should be exploited.

(iii) The concept of chemiclearance should be applied to packaging, since the relationship between polymer structure and resistance to radiation damage (including extractable products) can be established.

(iv) Approving a particular packaging material for use in a radiation sterilization procedure arrived at on the basis of extrapolation above the currently used dose is also possible. It can be done straightforwardly by referencing an established damage dose response relationship and extrapolating the packaging durability assessment indicates no compromise in safety or functionality, then the procedure can be considered acceptable.

## Packaging materials authorized for use for radiation-treated prepackaged food

| *No.* | *Packaging material* | *Max. dose (kGy)* | *Country*[b] | *Data*[b] |
|---|---|---|---|---|
| 1. | Cardboard | 10;35 | UK; Poland | 1991[c] |
| 2. | Polyethylene coextruded polyvinylacetate | 30 | USA; canada | 1988 |
| 3. | Polyethylene-co-vinylacetate | 30 | USA | 1989 |
| 4. | Fibreboard | 10 | India | 1997 |
| 5. | Fibreboard, wax coated (boxes) | 10 | USA; Canada | 1989 |
| 6. | Glassine paper | 10 | USA | 1975 |
| 7. | Glass | 10 | India | 1997 |
| 8. | Hessian sacks | 10 | UK | 1991[c] |
| 9. | Kraft paper | 0.5 | USA | 1975 |
| 10. | Nitrocellulose-coated cellophane | 10 | USA; India | 1975 |
| 11. | Nylon 11 | 10 | USA; India | 1975 |
| 12. | Nylon 6 | 60; 10 | USA; India | 1975 |
| 13. | Paper | 10; 35 | UK; Poland | 1991[c] |
| 14. | Paper coated or laminated with wax or polyethylene | 10; 35 | India; Poland | 1990 |
| 15. | Paper laminated with aluminium foil | 35 | Poland | 1990 |
| 16. | Polyamide film or polyamide coextruded with polyethylene | 35 | Poland | 1990 |
| 17. | Polyester-metallized-polyethylene laminate | 35 | Poland | 1990 |
| 18. | Polyester-polyethylene laminate | 35 | Poland | 1990 |
| 19. | Polyethylene film (various densities) | 60 ; 35, 10 | USA, Poland, India | 1975 |
| 20. | Polyethylene-paper-aluminium laminate | 35 | Poland | 1990 |
| 21. | Polyethylene-terephthalate | 60 | USA | 1975 |
| 22. | Polyolefin (low-density as middle or sealant layer) | | Canada | 1989 |
| 23. | Polyolefin (high-density as external layer) | | Canada | 1989 |
| 24. | Polyolefin film[c] | 10 | USA | 1975 |
| 25. | Polypropylene sacks | 10; 35 | UK; Poland | 1990[c] |
| 26. | Polypropylene – metallized | 35 | Poland | 1990 |
| 27. | Polystyrene film | 10 | USA; India | 1975 |

| No. | Packaging material | Max. dose (kGy) | Country[b] | Data[b] |
|---|---|---|---|---|
| 28. | Polystyrene foam trays (Styron 685 D) | 10 | Canada; India | 1975 |
| 29. | Rubber hydrochloride film | 10 | USA; India | 1975 |
| 30. | Steel, tin plated or enamel lined | 10 | India | 1997 |
| 31. | Vegetable parchment | 60; 10 | USA; India | 1975 |
| 32. | Vinylchloride-co-vinylacetate film | 60; 10 | USA; India | 1975 |
| 33. | Vinyldenechloride-coated cellophane | 10 | USA | 1975 |
| 34. | Vinylchloride-co-vinylideneechloride Film | 10 | USA; India | 1975 |
| 35. | Wood | 35; 10 | Poland; India | 1990 |
| 36. | Viscosa | 35 | Poland | 1990 |

a) Adapted from reference 466 with permission. Updated by Secretariat of the International Consultative Group on Food irradiation. September 1997.

b) Approvals USA – 1975 Canada – 1989; Poland – 35 kGy. 1986; United Kingdom – 1991; India – 10 kGy. 1996; earliest date of approval is cited

c) For dry herbs.

Chapter 15

# MARKETING OF IRRADIATED FOODS

PFA Rule 42 (zzz)(7) stipulates that "all packages of irradiated food shall bear the following declaration and logo,

PROCESSED BY IRRADIATION METHOD

DATE OF IRRADIATION—

Licence No.

Purpose of Irradiation

Like wise, as part of its approval, FDA requires that irradiated foods include labelling with either "treated with radiation" or "treated by irradiation" and the above international symbol for irradiation, the Radura."

In India, PFA Rule 49(26) stipulates:

*"Conditions for sale of irradiated food–*. All irradiated food shall be sold in prepacked condition only. The type of packaging material used for irradiated food for sale or for stock for sale or for exhibition for sale or for storage for sale shall confirm to the requirements of packaging material as per Rule 49(5)"

Paradoxically, Rule 48-D stipulates

"Storage and Sale of Irradiated food–Save as otherwise provided in these rules, no person shall Irradiate for sale, or transport for sale irradiated food."

## FDA stipulates

"Irradiation labelling requirements apply to only foods sold in stores. For example, irradiated spices or fresh straw berries should be labelled. When used as ingredients in other foods, however, the label of the other food does not need to describe these ingredients as irradiated. Irradiation labelling does not apply to restaurant foods also."

Despite the review in the May-June 1998 issue of "FDA Consumer" under the head *Irradiation–A safe Measure for Safer Food*" by John Henkel, quoting FDA scientists that irradiation reduces or eliminates pathogenic bacteria, insects and parasites; it reduces spoilage, and in certain fruits and vegetables, it inhibits sprouting and delays the ripening process. Also, it does not make food radioactive, compromise nutritional quality, or noticeably change food taste, texture or appearance as long as it is applied properly to a suitable product.

It continues:-

Health experts say that in addition to reducing *E.coli.0157-H7* contamination, irradiation can help control the potentially harmful bacteria *Salmonella* and *Campylobacter*, two chief causes of food-borne illness. In USA, the Centre for Disease Control and Prevention estimates that *Salmonella*-commonly found in poultry, eggs, meat and milk,–sickens as many as 4 million and kills 1,000 per year. *Campylobacter*, found mostly in poultry, is responsible for 6 million illnesses and 75 deaths/year in the United *States*. A May 1997 presidential report, "*Food safety from Farm to Table*" estimates that millions of Americans are stricken by food-borne illness each year and some 9,000, mostly the very young and elderly, die as a result.

Food scientists and experts emphasize that,

(i) irradiation is a useful tool for reducing food-borne disease risk, and

(ii) it complements, but doesn't replace, proper food handling practices by producers, processors, and consumers.

## Marketing

Despite all this stated above, retail outlets have been slow to carry irradiated foods. This, experts say, is partially due to fear in the retailers that consumers won't buy the products based on misgivings about radiation in general. This misgiving is due to the association of food irradiation with nuclear energy, a source that carries its share of negative perceptions. In USA, a Louis Harris poll released in 1986 found that 76% of Americans considered irradiated food a hazard. In 1995, researchers at the University of Georgeia reported that 87.5% of consumers had heard of irradiation but knew little about it. So the university set up a "*simulated super market setting*" and labelled irradiated products, put posters at the point of sale, and developed slide show explaining irradiation.

The goal was to see which one of these techniques was most effective in changing people's attitudes.

Kay Mc Walters, an agricultural research scientist and one of the study authors found in this study that any kind of education helps convey the benefits of irradiation.

But the one that turned out most effective was the slide show because the usual images and [narration] are much more attention-getting than just a static label or poster.

After the study's education strategy, about 84% of participating consumers said irradiation is "some what necessary" or "very necessary". 58% said they would always buy irradiated chicken if available, and 27% said they would buy it sometimes.

Another study in 1997 by the Food Marketing Institute had similar results. After receiving education about the process, 60% of those in the study said they would buy irradiated foods.

Consequently some stores have plunged in anyway with limited success. "Carrot Top", a Chicago-area grocery market owner Mr. Jim Conigan says that irradiated products have been selling steadily since 1992. He also discovered the above education effect on a small scale after sending his regular customers information about irradiation about irradiation in periodic news letters.

The layman gets scared even of the word "irradiation". He carries images of atomic explosions or nuclear reactor accidents. But if it is conveyed to him either through slides, or news letters that food irradiation is similar to sending luggage through an airport scanner, the process passes food quickly through a radiation field-typically gamma rays produced from radioactive $^{60}$Co. The amount of energy is not strong enough to add any radioactive material to the food like x-rays to the luggage.

The same irradiation process is used to sterilize medical syringes, bandages, contact lens solutions, gloves, sutures and gown. Like these irradiated medical products which do not transfer any radioactivity to the users, irradiated foods also behave similarly.

FDA has further affirmed that chemicals called radiolytic products created during food irradiation do not pose any health hazard. In fact, the same kinds of products are formed when food is cooked or barbecued. Along with the dissemination of above facts to consumers by audio visual systems, the food industry needs to get more irradiated products into the market place.

FDA's Dr. Gange Paul, Ph D. says that

"Most people in this country haven't even seen an irradiated food. When products start appearing, then the public can make up its mind"

## Arguments by Opponents

There are at least three prominent arguments against the irradiation of food, their rebuttal is also given alongwith.

(i)2-alkylacyclobutanones (2-ACB), which are unique to irradiated foods, are oncogenic and mutagenic in animals and are harmful to people who consume irradiated food. This claim refers to European research findings from 2002.[1,2] But these authors did not investigate the safety of irradiated foods but did report that formulations of chemically synthesized 2-ACBs, in concentrations about 1000 times those found in irradiated foods, had genotoxic and cytotoxic properties in vitro and that in rats treated with a known carcinogen, exposure to those concentrations of 2-ACB may promote the development of tumours. *The authors specifically cautioned against using their data to indict food irradiation.*

The European Commission's Scientific Committee on Food reviewed the research and, affirming its support of the WHO's assessment of irradiation safety, concluded that

"evidence of genotoxicity had not been established by standard methods and that the findings could not be considered relevant to the question of the safety of irradiated food products.[3]

Given the available evidence, any claim that the current studies of 2-ACBs are relevant to the safety irradiated foods is lacking in scientific credibility.

(ii) Irradiation destroys the nutritional quality of food is their second argument became the addition of any energy to food can break down its nutrients and molecules. Generally, macro molecules like fats, proteins and carbohydrates are not appreciably affected by irradiation. Thiamine (Vit $B_1$) is among the victims most sensitive to radiation, but food irradiation is not considered to threaten thiamine in the diet. The above review[4] by FDA and an independent Argentinian study[5] have concluded that irradiation poses no important risk to any nutrient in the diet, a conclusion supported by the American Dietetic Association.[6]

(iii) Their third argument is that irradiation is a quick fix and a technological solution to a policy problem. Food irradiation has been poutrayed as an easy way for industry and government to cover up or ignore the state of sanitation in processing facilities for meat and poultry. Traditional safety measures have the primary role is ensuring the safety of our meat supply, but they will not eliminate all contamination particularly in a slaughter house environment. For example, testing of *E. coli 0157:H7* in ground beef by the Deptt. of Agriculture's Food Safety and Inspection Service in 2003 showed that only 0.32% is contaminated[7]. In a country like USA even this small contamination means that an estimated 11.6 million kg of ground beef by the Deptt. of Agricultural's Food Safety and Inspection service is contaminated with *E.Coli 0157:H7.*

Irradiation cannot prevent primary contamination, but it can help ensure that contaminated ground beef does not reach the market place.

**Future opportunities**

As irradiated foods become widely available, public demand and public health advocacy groups will determine whether the irradiation of food will extend beyond its current niche to have a measurable effect on food safety. In the 1960s or 1970s physicians and allied health professionals had an important role in consumer's acceptance of the pasteurization of milk. As health advocates, they need to fill that role again in the adoption of food irradiation. It is important for physicians and other health professionals to be able to answer patient's questions accurately regarding the irradiation of food; to recommend irradiated foods, particularly for immuno compromised people, pregnant woman, children, and the elderly; to encourage local and state medical professional organization to endorse the use of irradiated products; to encourage grocery based departmental stores to stock irradiated foods; and to support the use of irradiated foods in school lunch programmes.

The European Commission (EC) adopted its first report in the European Union (EU). The report is based on the results of inspections carried out between September 2000 and December 2001 in Member states to determine the level of compliance with legislation governing food irradiation.

The commission expressed its satisfaction with the high level of compliance as only 1.4% of the 6,748 food samples tested across the EU were found to be irradiated and *labeled incorrectly* as tabulated below:-

**Summary of results for those Member States that carried out Surveillance.**

| *Member State* | *Results: Non-irradiated* | *Results: Irradiated and incorrectly Labelled* |
|---|---|---|
| Austria | 21 | 0 |
| Finland | 153 | 4 |
| Germany | 5491 | 26 |
| Greece | 99 | 0 |
| Ireland | 315 | 2 |
| Netherland | 88 | 0 |
| Sweden | 5 | 1 |
| United Kingdom | 479 | 64 |
| Total | 6651 | 97 |
| %of analyzed sample | 98.6 | 1.4 |

At the beginning of the reporting period, only facilities in the UK and the Netherlands were approved under 1999/2/EC while those in Belgium, Denmark, France and Germany were nationally authorized. However, all facilities were found to be in general compliance with the legislation and all are now authorized under EU legislation along with facilities in Spain. Results submitted from Ireland by the Food Safety Authority of Ireland (FSAI) were in line with general trends with just 2 out of 317 (0.6%) herbs and spices tested being found to be irradiated and not labelled properly.

Results from the UK however showed that 42% of certain dietary supplements tested were irradiated but not labelled appropriately. Because of the scale of the labelling problem identified by the UK and since most of those products cannot be legally irradiated in the EU, the Commission requested that all Member States carry out a survey of dietary supplements marketed within their own jurisdictions to determine the level of compliance with irradiation legislation.

In U.K., Food standard Agency is committed to ensuring that consumers are not misled about the food they buy. In the UK only correctly labelled irradiated herbs, spices or vegetable seasonings are permitted. Since 1996 several reports claimed that irradiated food, including dietary supplements and prawns and shrimps were in sale in the UK. As a follow up action, the Agency conducted a survey about food labelling in UK.

In U.K. the law states that if a permitted irradiated food is mixed with a non-irradiated food, the resulting product has to be labelled as either irradiated or treated with ionising radiation

The dietary supplements sampled alfalfa, *Aloevera*, cats, claw, devil's claw, garlic, ginger, *Ginkgo biloba*, ginseng, green tea, guarana, kava kava, saw palmetto, silymarin (milk thistle) and turmeric.

The spice products were aniseed, black pepper, cayenne pepper, cinnamon, chilli powder, cloves, coriander, cumin, curry paste, curry powder, ginger, *garam masala*, mixed spice, nutmeg, paprika, turmeric and white pepper. The dried herb products, sampled were : basil, dill, oregano, parsley, rosemary and thyme.

A range of product descriptions were found for prawns and shrimps including : black tiger, cap tiger, cock tail, cold water, crevettes, fantail, fantail tiger, fresh water, gaint, jumbo tiger king, king, *Macrobrachium*, Mekong tiger, North Atlantic, pink, salad, tiger king and tiger tails.

The above products were sampled during August and September 2001.

The survey results are:

Of 138 dietary supplement samples analyses, 44 (32%) had been wholly irradiated and a further 14 samples (10%) contained irradiated ingredients.

Of the 203 herb and spice samples, only one sample of ground nutmeg had an irradiated component.

Of the 202 prawn and shrimp samples analysed, only one sample was identified as having been irradiated. A further 4 samples were identified as containing irradiated components.

## INDIAN SCENARIO

In India, the rules 42(zzz)(7), 48-D, 49(26), 73 all were brought in finally wide GSR 614 (E) dated. 9.8.1994 (w.e.f. 9.8.1995) as corrected by GSR 60(E) dated. 7.2.195. Rice irradiation was allowed w.e.f. 6.4.1998 and fresh sea foods w.e.f. 2.5.2001.

Gamma irradiation facilities came into existence since 1974 when ISOMED irradiation plant was set up at Trombay by BRIT. It has a designed capacity to house 1000 kCi of $^{60}$Co, (1994). Subsequently RASHMI irradiation plant at Bangalore (1988) and SARC irradiation plant at Shiran Institute, Delhi (1990) were set up. These plants now have a design capacity of 300 kCi and 800 kCi respectively. RAVI irradiation plant at Jodhpur was set up by DRDL with a design capacity of 500 kCi $^{60}$Co in (1994). A commercial demonstration plant for radiation processing of spices was commissioned under BRIT at Navi Mumbai in the year 2000 with a designed capacity of 1000 kCi $^{60}$Co, KRUSHAK (Krishi Utpadan Sanrakshan Kendra) irradiation plant at lasalgaon, near Nashik, Maharashtra a 15 kCi $^{60}$Co $\gamma$-irradiation (earlier known as POTON irradiator) is a technology demonstration unit set up for processing onions for sprout control and other low dose applications of radiation for preservation of agriculture commodities. The installation and commissioning of various systems for KRUSHAK has been completed. Another plant in the private sector is under erection at Kundli near Delhi by Vardan Agro. for irradiation fruits and vegetables etc.

SARC (Delhi) plant has been licensed to irradiate spices only. It is occasionally used by spice manufactures for their export obligations so also the Navi Mumbai plant, which is located near the main spices export market. No plant has been licenced to irradiate meat and meat products including chicken, fresh/dried frozen sea food as well as fruits like mango, dried dates, raisins etc.

In other words, irradiation of permitted foods have yet to acquire take off stage, permission to irradiate more foods is a distant dreamed destination.

There is a tendency to put overseas consumer ahead of his domestic counterpart in providing quality food items. While it is true that expert markets are more lucrative and demanding but it dos not mean that quality requirements of domestic consumers require less efforts and attention. By definition and practice a world class food product shall have the same attributes both for domestic as well as overseas consumer. In fact, main reason for lack lustre performance in the quality front is the dual quality approach for different markets. While it makes business sense in many other industries to respond with different quality products to meet different markets need, food business has honourable exception as the value of human life is the same world over. Hence bringing in international quality standards to domestic market is one of the options available for improving quality (it has happened) though the process is tedious and expensive in the transformation stages.

Consumers awareness about quality of food products is mainly restricted to cities and other urban centres and it is yet to take roots in rural India, where a major portion of food is consumed.

Just like energy audits and productivity audits, quality audits should be made popular among food enterprises. Since making it mandatory again affect the spirit behind the concept, units coming forward for periodic audits both internal and external should be reimbursed the cost till the concept gains momentum.

The major function of quality control is to determine the durations of the product from the specification and when necessary make the needed changes to control the product quality at the desired level. The main role of the standards is to safeguard the health of the consumer and help him to get his money's worth and at the same time guide the processor or manufacturer to supply products acceptable to the consumer. They are aimed at ensuring the supply of pure, genuine and safe products. Quality control should cover all aspects, viz., physical, chemical, microbiological, toxicological, nutritional and organoleptic.

There are four common methods for arriving at the standards for product quality;

(i) legal standards
(ii) industry standards
(iii) consumer standards and
(iv) voluntary standards.

Following are the basis for determining product quality

A - Subjective and

B. Objective - i.e. physical, chemical and microscopic methods.

For food safety, food irradiation is the best. For this Rule 48-D needs to be abolished and India should follow the FDA rule, stated at the beginning of this chapter. To improve the administration of quality control for food products, a minimum infrastructure and tasting facilities should be established. This infrastructure shall ensure that a minimum sanitary and hygienic condition are there. The processing units failing to adhere to the prescribed norms may be debarred from production till the rectification of the pointed out defects.

## REFERENCES

1. Burnouf D, Delincee H, Hart wig A, et.al. "*Toxicological study to assess the risk associated with the Consumption of irradiated fat-containing food*". Serial Report INTERREG II Project no. 3,171. Karlsrule, Germany, Accessed April 8,2004.
2. Raul F, Gosse' F, Delincee H, *et al.* "*Food-borne radiolytic (2-alkylcyclobutanones may promote experimental colon carcinogenesis*" *Nutr Cancer*, (2002); 44,189-191.
3. Statement of the Scientific Committee on Food on a report on 2-alkylcyclobutanone. *Brussels*: European Commission, 3 - 7, 2002.
4. Food and Drug Administration. *Irradiation with Production and Handling of Food*: 21CFR part 179, Fed. Regist. 1997; 62: 64107-64107.
5. Narvaiz P, Ladomery L.; *Estimation of the effect of food irradiation on total dietary intake of vitamins as compared with dietary allowance*: "*Study for Argentina*". *Radiat. Phys. Chem.*, 1996, 48, 360-1.
6. Wood OB, Bruhn CM. *Position of the American Dietetic Assn: Food Irradiation, J. Am. Diet. Assoc.* 2000, 100, 246-253.
7. Roybal J. *Beef industry logs successful week in E.coli. 0157: H7 battle; Beef Magazine's Cow-Calf Weekly*. Sept. 26,2003.

Chapter 16

# ATOMIC ENERGY (CONTROL OF IRRADIATION OF FOOD) RULES, 1996

## CONTENTS

19. Monitoring of personnel
20. Operating procedures
21. Emergency procedures
22. Security of irradiation facility
23. Posting of radiation warning signs
24. Radiation survey instruments
25. Decommissioning
26. Disposal of decayed sources

| | | |
|---|---|---|
| Schedule I | - | Technological conditions |
| Schedule II | - | Qualifications of personnel |
| Schedule III | - | Document to be submitted to the competent authority for obtaining approval of irradiation facilities |
| Schedule IV | | See rules 3 and 6 |
| Schedule V | - | General conditions for Irradiated Foods |
| Schedule VI | - | Radiation Survey Instruments required for radiation monitoring. |
| Schedule VII | - | Standards of Packaging Material. |
| Schedule VIII | - | Operational Limits |
| Schedule IX | - | Leakage and safety related tests |
| Form I | - | Application for Obtaining a Licenser for an Irradiation facility |
| Form II | - | Certificate of approval for irradiation facility |
| Form III | - | Application for obtaining the certificate of approval for an irradiation Facility |
| Form IV | - | Licence for Operating An Irradiation Facility |
| Form V | - | Record of food irradiation |

## Atomic Energy (Control of Irradiation of Food) Rules, 1996

*G.S.R. 254–Mumbai, the 29th March, 1996, in exercise of the powers conferred by section 30 read with sections 14 and 17 of the Atomic Energy Act, 1962 (33 of 1962) and in supersession of the Atomic Energy (Control of Irradiation of Food), Rules, 1990, except as respects things done or omitted to be done before such supersession, the Central Government hereby makes the following rules, namely:-

1. **Short title and commencement**–(1) Theses rules may be called the Atomic Energy (Control of Irradiation of Food) Rules, 1996.

   (2) They shall come into force on the date of their publication in the Official Gazette.

2. **Definitions**–In these rules unless the context otherwise requires:-
   - (a) "Act" means the Atomic Energy Act, 1962 (33 of 1962 ),
   - (b) "applicant" means the person making an application for the grant of licence;
   - (c) "Certificate of approval" means the certificate of approval granted by the competent authority for operating an irradiation facility in accordance with rule 5;
   - (d) "Certificate of irradiation" means the certificate of irradiation issued by the licensee under rule 12;
   - (e) "Certificate of release" means certificate of release issued by the licensing authority under rule 25;
   - (f) "Competent authority" means such officer or authority directed by the Central Government under Section 27 of the Act to exercise the powers or discharge the duties under these rules;
   - (g) "dosimetry" means the method to measure the absorbed dose of radiation by the food;
   - (h) "food" means article of food referred to in column 2 of Schedule I;
   - (i) "Form" means the Form annexed to these rules;
   - (j) "inspection report" means the inspection report referred to in rule 10;

(k) "irradiation facility" means any facility which is capable of being utilised for the treatment of food by radiation;

(l) "irradiation food" means articles of food subjected to radiation by:–

(i) Gamma rays;

(ii) X-rays generated from machine sources operated at or below an energy level of 5 million electron volts;

(iii) Sub-atomic particles, namely electrons generated from machine sources operated at or below an energy level of 10 Million electron volts, to dose levels as specified in Schedule I.

(m) "licencee" means licence for operating an irradiation facility issued under rule 4;

(n) "licence" means the person named in the licence in whose favour the licence is issued;

(o) "Licensing authority" means the Central Government or such officer or authority directed by the Central Government under section 27 of the Act to grant licence under section 14 of the Act;

(p) "operational limits" means limits specified by the competent authority on the level of radiation to workers and members of the public and on the level of radioactive contamination from the radioactive sources used in the irradiation facility;

(q) "operator" means any person, appointed as such by the licensee and who shall have the qualification and other requirements as specified in paragraph 2 of Schedule II;

(r) "Quality Control Officer" means any person appointed as such by the licensee who shall have the qualification and other requirements as specified in paragraph 3 of Schedule II;

(s) "Radiological Safety Officer" means any person appointed as such by the licensee who have the qualification and other requirements as specified in paragraph 1 of Schedule II;

(t) "Schedule" means Schedule annexed to these rules.

3. **Conditions precedent for the issue of a licence**– No licence for operating an irradiation facility shall be granted, unless the applicant obtains a certificate of approval from the competent authority, after submitting documents as specified in Schedule III to the effect that the facility is in conformity with the general conditions for design, operation and efficiency criteria as specified in Schedule IV.

4. **Licence**– (1) An application for a licence shall be made to the licensing authority in Form I and shall be accompanied by:–
   (a) a certificate of approval;
   (b) fee of rupees five hundred payable in the form of a bank draft drawn in favour of the licensing authority.

2. If the licensing authority is satisfied that the applicant is capable of operating an irradiation facility in accordance with these rules, it shall issue a licence in Form IV.

3. If the licensing authority is satisfied that the applicant is not capable of operating an irradiation facility in accordance with these rules, it may, after giving the applicant reasonable opportunity of being heard against the proposed refusal of licence, by order setting out the reasons therein, refuse to grant the license.

4. Every order refusing to grant the licence sub-rule (3), shall be communicated to the applicant by sending a copy of the order by registered post to the address given in the application.

5. **Certificate of approval**– (1) An application for a certificate of approval shall be made to the competent authority in Form III.

   (2) If the competent authority is satisfied that the applicant satisfies the requirements as specified in Schedule III and IV, it shall grant a certificate of approval in Form II.

   (3) If the competent authority is satisfied that the applicant does not satisfy the requirements as specified in Schedule III and IV, it may, after giving the applicant a reasonable opportunity of being heard against the proposed refusal of the certificate of approval, by order, setting out the reasons therein, refuse to grant the certificate of approval.

(4) Every order refusing to grant the certificate of approval under sub-rule (3), shall be communicated to the applicant by sending a copy of the order by registered post to the address given in the application.

6. **Power to suspend the certificate of approval**– (1) If the competent authority is satisfied, on the basis of the inspection report, that any irradiation facility has ceased to conform to the safety and efficiency criteria as specified in Schedule IV; it may for reasons to be recorded in writing, make and other suspending the certificate of approval and call upon the licensee to rectify the defects mentioned therein within a period of thirty days from the date of receipt of the order.

(2) The order of the competent authority made under sub-rule (1) along with the copy of the inspection report shall be communicated to the licensee who shall on receipt of the same, cease to operate the irradiation facility until the suspension of the certificate of approval is revoked under sub-rule (5).

(3) The competent authority shall transmit copies of the order and inspection report referred to sub-rule (2) to the licensing authority.

(4) The competent authority shall enter in the certificate of approval the particulars of inspection.

(5) Where the competent authority is satisfied that pursuant to the order made under sub-rule (1) the licensee has rectified the defects mentioned therein, it may revoke the suspension of the certificate of approval and communicate in writing the decision to the licensee.

(6) If the defects mentioned in the order are not rectified by the licensee within the period mentioned therein, the competent authority shall report the matter to the matter to the licensing authority for necessary action.

(7) Any person aggrieved by the order of the suspension of certificate of approval by the competent authority may within a period of fifteen days from the receipt of the communication of such suspension prefer an appeal to the Atomic Energy Commission.

7. **Power to suspend the Licence**–(1) If the licensing authority is satisfied, on the basis of the inspection report that the licensee has failed to ensure that the irradiation facility under his control conforms to the safety and efficiency criteria specified in Schedule IV, it may for reasons to be recorded in writing make an order suspending the license and call upon the licensee to rectify the defects mentioned therein within a period of thirty days from the date of receipt of the order.

   (2) The period of thirty days stipulated under sub-rule (1) of rules 6 and 7 may be condoned by the licensing authority after having satisfied himself with the fact that the licensee is not able to rectify the defects for reasons beyond his control within the stipulated period of thirty days.

   (3) The order made by the licensing authority under sub-rule (1) along with the copy of the inspection report shall be communicated to the licensee, who shall on receipt of the same, cease to operate the irradiation facility until the suspension of the license is revoked.

   (4) Where the licensing authority is satisfied that pursuant to the order made under sub-rule (1), the licensee has rectified the defects mentioned therein, it may revoke the suspension of the licence and communicate the decision to the licensee.

   (5) Any person aggrieved by the order of suspension or revocation of a licence by the licensing authority may within a period of fifteen days from the receipt of the communication of such suspension or revocation prefer an appeal to the Central Government.

8. **Revocation of licence**– Notwithstanding anything contained in rule 4, the licensing authority may after giving the licensee a reasonable opportunity of being heard, by order setting out the reasons therein in writing, revoke any licence or modify the terms and conditions of any licence on any of the following grounds, namely:–

   (1) Where the certificate of approval has been suspended and the competent authority has reported that the licensee has not rectified the defects mentioned in the order of suspension within the period stipulated therein; or

(2) Where the licence has been suspended and the licensee has not rectified the defects mentioned in the order of suspension within the period stipulated therein; or

(3) Where in the opinion of the licensing authority, it is necessary to do so, in order to ensure safety of persons whosoever may suffer any injury because of any negligence in the process of irradiation.

9. **Conditions for irradiation of food**– (1) The licensee shall not undertake irradiation of any food unless in his opinion such irradiation is necessary for its preservation, protection against parasites or improvement of its hygienic or technological quality.

(2) the licenses shall ensure that :–

(a) the Quality Control Officer has satisfied himself that the food to be irradiated is of good quality,

(b) in the case of packaged products, the packing material conforms to the standards specified in Schedule VII,

(c) the irradiated food is readily identified so as to prevent it from being subjected to subsequent irradiation,

(d) the dose limit, radiation source and irradiation conforms to the conditions specified in Schedules I and V,

(e) the irradiation facility is operated only by the operator.

10. **Periodic Inspection of facilities**— (1) The competent authority or any person authorized by him in this behalf shall undertake inspection of the irradiation facility at least twice in a year but the maximum gap between two inspections shall not exceed 8 months.

(2) The particulars of the inspection shall be recorded by an entry on the certificate of approval.

(3) A copy each of the inspection report shall be forwarded to the licensee and to the licensing authority.

11. **Record of food irradiation**—(1) The licensee shall maintain, for each source of radiation used, a record in Form V indicating, for each batch of food subjected to radiation treatment:–

(a) the serial number of the batch;

(b) the date of irradiation;

(c) the nature and the quality of irradiation treatment in the case of packaged products;

(d) the type of packaging used during the radiation treatment in the case of packaged products;

(e) the control and measurement performed during the treatment, particularly as regards the minimum and maximum limits of radiation dose;

(f) where appropriate, all supplementary information required by the specific irradiation conditions provided for in Schedule V;

(g) any incidents and anomalies observed during the irradiation treatment.

(2) The record shall contain the names, and addresses of the operators and the quality control officer and the identification number of the irradiation facility.

(3) The licensee shall retain the records for a period of five years.

(4) The licensee shall maintain the records of standard model for the food irradiation in respect of each batch of food as specified in Form V.

**12. Certificate of irradiation**– (1) The licensee shall, on the basis of the data entered in the record of irradiated food, issue a certificate of irradiation for every batch of food item which has undergone irradiation in his facility.

(2) A copy of the certificate of irradiation shall be maintained by the licensee.

**13. Irradiation voucher**— The licensee shall issue to the person from whom an order for the irradiation of food has been received, an irradiation voucher for each batch of food containing the following particulars, namely :–

(1) the identification number of the irradiation facility together with the names and addresses of the operators and quality control officer;

(2) the nature and quality of the batch of irradiated food and also the purpose of the irradiation;

(3) the date of the radiation treatment;

(4) the radiation source used and specific dose of radiation;

(5) the serial number of the batch which has been subjected to the treatment which number must correspond to the information in the irradiation record;

(6) date and signature of the licensee.

**14. Maintenance of log books**– Every licensee shall maintain and make available to any person duly authorized by the competent authority for inspection, a log book containing the following particulars, namely :–

(1) description of the facility;

(2) source details;

(3) the name of supplier of the source and his address;

(4) the identity of Radiological Safety Officer (RSO).

**15. Control of irradiation facility**– (1) The competent authority or any person authorized by him in this behalf shall undertake verification to ensure that the operation of the facility and the use of radiation treatment procedures conform to the general and specific conditions as specified in Schedule IV.

(2) The licensee shall provide all reasonable facilities to the competent authority or any person authorized by him, to carry out inspection and measurement procedures as may be necessary.

**16. Powers and duties of person authorized to inspect**– (1) The competent authority or any person authorised by him in this behalf shall–

(a) have the right of access :–

(i) to any place which is used for irradiation of food or for storage of food which has been or has to be irradiated;

(ii) to all documents relating to the irradiation facility, the batches of food which have been or are to be irradiated a certificate of approval, the copies of the inspection report, order of suspension, modification or revocation of licence, food radiation records and the documents relating to purchase and sale accompanying the batches of food.

(b) to check the performance of their irradiation unit and measure the dosage to which the food is subject to.

**17. Radiological safety officer**-- The qualifications for Radiological Safety Officer are given in Schedule II. No person shall be appointed as a Radiological Safety Officer unless such person :–

(1) possesses a certificate towards his qualification by an institution or authority approved by the competent authority.

(2) is familiar with the operating and emergency procedures of the irradiation unit and has demonstrated his understanding thereof to the satisfaction of the competent authority.

(3) is familiar with the rules, notification and orders applicable to the irradiation facility and has demonstrated his understanding thereof to the satisfaction of the competent authority.

(4) has demonstrated to the satisfaction of the Competent authority competence to use the irradiation facility and the radiation instruments to be used in his assignment.

**18. Duties of Radiological Safety Officer**– The Radiological Safety Officer shall–

(1) instruct the radiation workers under his charge on the hazards of radiation and on suitable safety measures and work practices aimed at minimizing exposure to radiation;

(2) take all necessary steps aimed at ensuring that the operational limits as specified in Schedule-VIII are not exceeded;

(3) carry out leakage tests as specified by the competent authority in Schedule-IX and test of safety related systems;

(4) investigate and initiate prompt and suitable remedial measures in respect of any situation that could result in radiation hazard;

(5) ensure that reports on all hazardous situations along with details of any immediate remedial measures that may have been initiated, are made available immediately to his employer;

(6) assist the licensee in the safe disposal of decayed radioactive sources in the manners approved by the competent authority.

**19. Monitoring of personnel**– The licensee shall ensure that:–

(1) every person entering the irradiation facility wears separate personnel monitoring device provided by him.

(2) such personnel monitoring devices are processed by an agency approved by the competent authority.

(3) Investigation reports regarding all excessive exposures are forwarded to the competent authority.

(4) The workers exposed to radiation have been medically examined as specified by the competent authority.

**20. Operating procedures**– The licensee shall prepare a detailed operating procedure based on the manufacturer's manual which shall inter alia provide that:–

(1) the irradiation facility shall be operated only in such a way that no person is likely to be exposed to radiation doses in excess of the operational limits notified by the competent authority.

(2) it shall be impossible to resume operation of irradiation facility after the return of the source to the fully shielded position without complying with the requirements such as interlocked controls for personnel access, radiation room lock-up sequence and source exposing operations.

(3) a single multipurpose key which is to operate the control console, to gain access to the radiation room and to activate safety related interlocks is available.

(4) the procedures and occasions for conducting radiation surveys and contamination tests, are devised.

(5) the procedures for locking and the use of facility are available.

(6) monitoring of the personnel and the use of area monitoring equipment is made mandatory.

(7) transportation of sealed sources used in irradiation unit is undertaken in a safe manner.

(8) records are maintained in proper from.

(9) for inspection and maintenance procedures, approved by the competent authority, are devised for the radiation facility.

(10) in the event of an accident persons to be notified in accordance with the procedure specified by the competent authority.

(11) the radiation zone monitor is so integrated with the personnel access door interlocks to prevent room access of specified values of if the detector malfunctions or is turned off.

(12) the monitor shall also generate audible and visible alarm signals if the radiation level exceeds that when source is in the fully shielded position.

**21. Emergency procedures**— (1) The licensee shall prepare detailed emergency procedures for each type of emergency that may reasonably be anticipated.

(2) The procedure should be brief and should be expressed in the form of instructions which can be understood by non-technical persons.

(3) The procedures should describe situations requiring emergency action and specify immediate action to be taken to minimise radiation dose to persons in the vicinity of the irradiation facility.

**22. Security of irradiation facility**– The licensee shall ensure that during the operation of the irradiation facility, direct surveillance of the facility is maintained in order to prevent unauthorized entry Licensee shall conspicuously display the radiation unto the facility.

**23. Posting of radiation warning signs**– The warning sign as given below at the site in which irradiation is in progress.

**24. Radiation survey instruments**– The licensee shall not commence operation of any irradiation facility under his control unless the irradiation facility is equipped with instruments as specified by the competent authority in Schedule VI.

**25. Decommissioning**– (1) No licensee shall decommission any irradiation facility under his control unless he:–

(a) obtains the previous permission of the licensing authority;

(b) undertakes to decommission the irradiation facility in the manner specified by the competent authority;

(c) undertakes to bear all the expenses of such decommissioning.

(2) The licensee shall not use the site used for the irradiation facility for any other purpose, unless he obtains the certificate of release from the licensing authority.

(3) The licensing authority may on the application of the licensee and after obtaining a report from the competent authority, in its discretion issue a certificate of release.

**26. Disposal of decayed sources**– The licensee shall at his expense, ensure the safe disposal of the radioactive sources in such manner as may be approved by the competent authority.

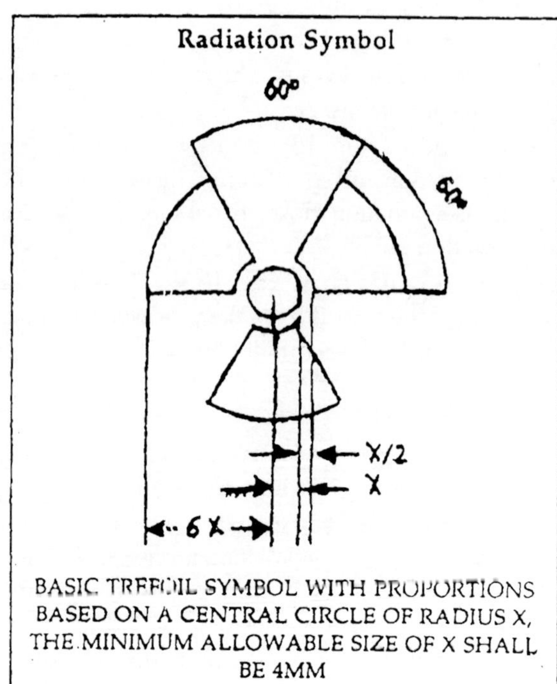


Radiation Symbol

BASIC TREFOIL SYMBOL WITH PROPORTIONS BASED ON A CENTRAL CIRCLE OF RADIUS X, THE MINIMUM ALLOWABLE SIZE OF X SHALL BE 4MM

## SCHEDULE I

[See rules 2(h) and 9(2) (d) ]

TECHNOLOGICAL CONDITIONS

| *Sl. No.* | *Name of food* | *Purpose of irradiation* | *Dose (kGy)* | | | |
|---|---|---|---|---|---|---|
| | | | *Minimum* | *Maximum* | *Overall average* | *Specific conditions* |
| *(1)* | *(2)* | *(3)* | *(4)* | *(5)* | *(6)* | *(7)* |
| 1. | Onions | To inhibit sprouting | 0.03 (30 Gy) | 0.09 (90 Gy) | 0.06 (60 Gy) | |
| 2. | Potatoes | To inhibit sprouting | 0.06 (60 Gy) | 0.15 (150 Gy) | 0.10 (100 Gy) | |
| 3. | Frozen sea-foods | To reduce the number of certain pathogenic microorganisms such as Salmonella in packaged frozen seafoods. | 4 | 6 | 5 | Irradiation should be carried out in a frozen state |
| 4. | Spices | To control insect infestation to reduce microbial load and pathogenic microorganisms | 6 | 14 | 10 | Irradiation under prepacked condition. |
| 5. | Rice | To control insect infestation | 0.25 | 1.0 | 0.62 | Irradiation under prepacked condition |
| 6. | Semolina (Sooji or Rawa), Wheat Atta and Maida | To control insect infestation | 0.25 | 1.0 | 0.62 | Irradiation under prepacked condition |
| 7. | Mango | To improve shelf-life and for quarantine purposes | 0.25 | 0.75 | 0.50 | – |
| 8. | Raisins, Figs, and dried Dates | To control insect infestation | 0.25 | 0.75 | 0.50 | Irradiation under prepacked condition |
| 9. | Ginger, Garlic and Shallots (Small onions). | To inhibit sprouting | 0.03 | 0.15 | 0.09 | – |
| 10. | Meat and Meat products including Chicken | To reduce number of spoilage micro-organisms and certain pathogenic microorangisms and parasites | 2.5 | 4.0 | 3.25 | Irradiation under prepacked condition and to be carried at 0-3°C.] |

## SCHEDULE II

[See rule 2(q), (r) and (s)]

QUALIFICATIONS OF PERSONNEL

1. **Radiological Safety Officer**– The minimum qualification : He should be a Science graduate with physics as one of the subjects.

He should have also successfully undergone instructions specified for the training of Radiological Safety Officer level III as prescribed by the Atomic Energy Regulatory Board and should posses a valid certificate to that effect. The subject for instructions include fundamentals of radiation and radiation protection, concept of dose limit, use of instruments and survey techniques in radiation detection; and knowledge of personnel monitoring equipments, inspection and maintenance of safety interlocks, and operation and emergency procedures.

2. **Operator**– The operator should be Science graduate/ Diploma holder either in Electrical or Mechanical Engineering.

The subject for inclusion in the instruction of the operator will be the same as those prescribed for Radiological Safety Officer except that he shall be given training in elements of food technology and food irradiation technology also.

Food Irradiation Technology :–

(a) Details of radiation sources used for food irradiation.
(b) Measurement of absorbed doses during irradiation using proper dosimeter for various types of sources and maintenance of dosimetry and source activity.
(c) Source-product geometry to increase efficiency utilization.
(d) Operation of radionuclide irradiation to ensure correct operational and of correct safe position of the source which should be interlocked with the product movement.
(e) Maintenance of irradiator.
(f) Irradiation technology of individual food items.

3. **Quality Control Officer**– The qualification of the Quality Control Officer should be either M.Sc./B. Tech. in Microbiology, Food Technology or Food Chemistry.

(1) The Quality Control Officer shall ascertain the quality of the food.

## SCHEDULE III

(see rule 3)

## DOCUMENT TO BE SUBMITTED TO THE COMPETENT AUTHORITY FOR OBTAINING APPROVAL OF IRRADIATION FACILITIES

The applicant shall submit to the competent authority :–

(1) The complete design drawings of the facility indicating the details of the shielding surrounding the source, wall thickness, labyrinth access, openings voids, reinforcements, mechanical and electrical safety system, ventilation, fire protection systems.

(2) The environs of the facility including residential complexes, occupancy within 50 m radius of the facility, geology of the location, water table, soil characteristics, seismicity.

(3) Complete description of the radiation source, address of the supplier and the operating condition of the source such as source drive system.

(4) Safety analysis report to demonstrate the adequacy of radiation safety under normal and anticipated accident conditions.

## SCHEDULE IV

(See rules 3 and 6)

### A. Conditions for the operation of irradiation facilities

1. *Introduction* :–

   1.1 Only irradiation facilities based on the use of either a radionuclide source ($^{60}$Co or $^{137}$Cs) or X-rays and electrons generated from machine sources should be used.

   1.2 The irradiation facility, may be of two designs either "continuous" or "batch" type.

   1.3 The irradiation facility shall make use of accepted methods for measuring the absorbed radiation dose and of the monitoring of the physical parameters of the process.

2. *Irradiation plants:*–

   2.1 The manufacturer of the facility should state the activity of the source in Becqueral (Bq) in the case of radionuclide source, and it should be recorded. The recorded activity should take into account the natural decay rate of the source and should be accompanied by a record of the date of measurements or recalculation.

2.2 The irradiation facility using radionuclide sources shall have a well separated and shielded depository for the source element and a treatment area which can be entered only when the source is in the safe position.

2.3 There should be a positive indication of the correct operational and of the correct safe position of the same which should be interlocked with the product movement system.

2.4 In the case of machine sources the average beam power should be adequately recorded.

2.5 There should be a positive indication of the correct setting of all machine parameters which should be interlocked with the product movement system.

2.6 A beam scanner or a scattering device (e.g. the converting target) must be incorporated in a machine source to obtain an even distribution of the radiation over the surface of the product.

2.7 The product movement, with width and speed of the scan and the beam pulse frequency (if applicable) should be adjusted to ensure a uniform surface dose.

3. *Dosimetry and Control:–*

(1) Certain dosimetry measurements (para. B) should be made prior to the irradiation of any food stuff.

(2) The dosimetry measurement should be done for each new food irradiated, irradiation process, and whenever modifications are made to source strength or type and to the source product geometry.

(3) Routine dosimetry should be made during operation and records of such measurements must be available for inspection.

(4) The measurements of facility parameters governing the process such as transportation speed, dwell time, source

exposure time, machine beam parameters, should be made regularly during the operation and record must be kept for inspection.

4. *Good radiation processing practice:–*
   (1) The design of the irradiation should have the facility to optimize the dose uniformity ratio to ensure appropriate dose rates and where necessary to permit temperature control during irradiation (e.g. for treatment of frozen food) and also control of atmosphere.
   (2) Care must be exercised to minimize mechanical damage to the product during transport, irradiation and storage and to use irradiation to its maximum efficiency.
   (3) Where the food to be irradiated is subjected to special standards for hygiene or temperature control, the facility must permit compliance with these standards.

5. *Product and Inventory Control :–*
   (1) The incoming product should be physically separated from the outgoing irradiated product.
   (2) Where appropriate, a visual colour change radiation indicator should be affixed to each product pack for ready identification of irradiated and non-irradiated products.
   (3) Records should be kept in the facility record book which show the nature and kind of the product being treated, its identifying marks of packed or, if not, the shipping details, its bulk density the type of source or electron machine, the dosimetry used and details, of their calibration, and the dates of the treatments.
   (4) All products shall be handled, before and after irradiation, according to accepted good manufacturing practices taking into account of the particular requirements of the technology or the process (Schedule I). Suitable facilities for refrigerated storage may be required.

## B. Dosimetry

1. *The overall average absorbed dose :*

It can be assumed for the purpose of the determination of the wholesomeness of food treated with an overall average dose of 10 kGy or less, that all radiation chemical effects in that particular dose range are proportional to dose.

The overall average dose, D, is defined by the following integral over the total volume of the goods.

$$D=1/M \int [(x, y, z), d\ (x, y, z)]\ .\ dv$$

Where,

M = the total mass of the treated sample

$\int$ = The local density at the point (x, y, z)

d = the local absorbed dose at the point (x, y, z)

dv = dx dy dz the infinitesimal volume element which in real case is represented by the volume fractions.

The overall average absorbed dose can be determined directly by homogenous products or for bulk goods of homogeneous bulk density by distributing an adequate number of dose meters strategically and at random throughout the volume of the goods. From the dose distribution determined in the this manner an average can be calculated which is the overall average absorbed dose.

If the shape of the dose distribution curve through the product is well determined the positions of minimum and maximum dose are known. Measurements of the distribution of dose in these two position in a series of samples of the product can be used to give an estimate of the overall average dose.

In some cases the mean value of the average values of the minimum (Dmin) and maximum (Dmax) dose will be a good estimate of the overall average dose, i.e. in these cases.

$$\text{Over average dose} = \frac{\text{Dmax} + \text{Dmin}}{2}$$

2. *Effective and limiting dose values*

(1) Some effective treatment e.g. the elimination of harmful micro organisms or a particular shelf-life extension, or a

disinfestations requires a minimum absorbed dose. For other applications too high an absorbed dose may cause undesirable effects or an impairment of the quality of the product.

(2) The design of the facility and the operational parameters have to take into account minimum and maximum dose values required by the process. In some low dose applications it will be possible within the terms of paragraph 4 on Good Radiation Processing Practice to allow a ratio of maximum to minimum dose of greater than 3.

(3) With regards to the maximum dose value under acceptable wholesomeness considerations and because of the statistical distribution of the dose of mass fraction of product of atleast 95% should receive an absorbed dose of less than 15 kGy when the overall average dose is 10 kGy.

3. *Routine dosimetry*

Measurements of the dose in a reference position can be made occasionally throughout the process. The association between the dose in the reference position and the overall average dose must be known. These measurements should be used to ensure the correct operation of the process. A recognized and calibrated system of dosimetry should be used.

A complete record of all dosimetry measurements including calibration must be kept.

4. *Process Control*

(1) In the case of a continuous irradiation facility employing radionuclide, it will be possible to make automatically a record of transportation speed or dwell time together with indications of source and product positioning. These measurements can be used to provide a continuous control of the process in support of routine dosimetry measurements.

(2) In a batch operated radionuclide facility automatic recording of source exposure time can be made and a

record of product movement and placement can be kept to provide a control of the process in support of routine dosimetry measurements.

(3) In a machine facility, a continuous record, of beam parameters, e.g. voltage, current, scan speed, scan width, pulse repetition and a record of transportation speed through the beam can be used to provide a continuous control of the process in support of routine dosimetry measurements.

## SCHEDULE V

[See rule 9(2) (d)]

GENERAL CONDITIONS FOR IRRADIATED FOODS

1. *Scope*– These conditions apply to foods processed by irradiation. It does not apply to foods exposed to doses imparted by measuring instruments used for inspection purposes.
2. *General conditions for the process*– (1) *Radiation Source:*

The following types of ionizing radiation may be used :–

(a) Gamma rays from the radionuclide $^{60}Co$ or $^{137}Cs$;

(b) X-rays generated from machine sources operated at or below an energy level of 5 MeV;

(c) Electrons generated from machine sources operated at or below an energy level of 10 MeV.

*(1) Absorbed Dose*– The overall average dose absorbed by a food subjected to radiation processing should not exceed 10 kGy.

3. ***Facility and Control of the process.***

3.1. Radiation treatment of foods shall be carried out in facilities licensed and registered for this purpose by the licensing authority.

3.2. The facilities shall be designed to meet the requirements of safety, efficacy and good hygienic practices of food processing.

3.3. The facilities shall be staffed by adequately trained and competent personnel.

3.4. Control of the process with the facility shall include the keeping of adequate records including quantitative dosimetry.

3.5. Premises and records shall be open to inspection by appropriate national authorities.

4. ***Technological Requirements***– (1) *Conditions for Irradiation*– The irradiation of food is justified only when it fulfills a technological need.

(2) *Food Quality and Packaging Requirements*– The doses applied shall be commensurate with technological and public health purposes to be achieved and shall be in accordance with good radiation processing practice. Foods to be irradiated and their packaging materials shall be of suitable quality, acceptable hygienic condition and appropriate for this purpose and shall be handled, before and after irradiation, according to good manufacturing practices taking into account the particular requirements of the technology of the process.

5. ***Labelling***– (1) *Inventory control*– For irradiated foods, whether prepacked or not, the relevant shipping documents shall give appropriate information to identify the registered facility which has irradiated the food, the date(s) of treatment and lot identification.

(2) *Prepacked foods intended for direct consumption*– The labelling of prepackaged irradiated foods shall be in accordance with the relevant provisions given in Prevention of Food Adulteration Rules.

(3) *Food in Bulk containers*– The declaration of the fact of irradiation shall be made clear on the relevant shipping documents.

6. ***Re-Irradiation***– (1) Except for foods with low moisture content (cereals, pulses, dehydrated foods and other such commodities) irradiated for the purpose of controlling insect reinfestation, foods irradiated in accordance with 2 and 3 of this standard shall not be re-irradiated.

(2) For the purpose of this standard food is not considered as having re-irradiated when

(a) the food prepared from materials which have been irradiated at low levels e.g. about 1 kGy, is irradiated for another technological purpose;

(b) the food containing less than 5% of irradiated ingredient, is irradiated, or when

(c) the full dose of ionizing radiation required to achieve the desired effect is applied to the food is more than one instalment as part of processing for a specific technological purpose.

(3) The cumulative overall average dose absorbed should not exceed 10 kGy as a result of irradiation.

## SCHEDULE VI

[See Rule 24]

RADIATION SURVEY INSTRUMENTS REQUIRED FOR RADIATION MONITORING.

Either

A. 1. Radiation Survey Meter (G.M. Type 0.2 mR/hr. or with other appropriate measurement range.

and

2. Radiation Survey Meter (Ionization chamber type 0-5R/hr. or with other appropriate measurement range).

or

B. 1. Wide range Survey Meter (G.M. Type 0-100 R/hr. or with other appropriate measurement range).

2. Alternatively, any single or combination of radiation survey meters recognized by the competent authority as equivalent to the above.

## SCHEDULE VII

[See Rule 9(1) (b)]

STANDARDS OF PACKAGING MATERIAL

1. (1) Where packing is essential to prevent post-treatment recontamination, the food should be packed before treatment.

(2) Package size especially with bulk should be such that they can be handled efficiently thereby avoiding excessive delays and temperature abuse.

(3) The choice of the packing material and the nature of the container for specific food is usually determined by the purpose they are to serve and storage conditions, such a prevention of moisture loss or moisture uptake, to provide an atmosphere devoid of air, or to avoid mechanical damage to food.

(4) Sterilized food must have containers which prevent access to bacteria or other microorganisms.

2. *Rigid Containers*– (1) Primary rigid container used for irradiated food is metal can, steel containers, tin-plated and lined with appropriated enamel such as polybutadiene or epoxy-phenolic have been found to be satisfactory.

(2) Secondary rigid containers made of wood, fibre board or glass are also used.

3. *Flexible packaging material*– Flexible plastic containers because of their low density are highly suitable for packing irradiated food. Films over 76 cm thick are satisfactory for preservation of irradiated products. In case laminates are used their performance should satisfy the following criteria:–

(i) The films are not changed adversely :–

(a) in their protective characteristics (e.g. seal stability, permeability, etc.)

(b) by radiation induced changes in the food,

(c) causing transmission of toxic or potentially toxic, and substances to the food.

(ii) Polymeric films recommended for use upto 10kGy are:–

(a) Nitrocellulose or vinylidene coated cellophane;

(b) Wax coated paper board;

(c) Glassine paper;

(d) Polystyrene;

(e) Rubber hydrochloride;

(f) Vinylidene chloride – vinyl chloride;

(g) Polyethylene;

(h) Polyethylene terephthalate;
(i) Nylon 6;
(j) Nylon 11;
(k) Vinyl chloride –vinyl acetate;
(l) Vegetable parchment.

## SCHEDULE VIII

[See Rule 18(2)]

## OPERATIONAL LIMITS

### Effective dose limits

1. The cumulative effective dose constraint for five years from January 1, 1994, to December, 31, 1998 will be one hundred milli Sievert (100 mSv) for individual radiation workers.
2. The annual effective dose to individual workers in any calendar year during the five-year block shall not exceed the limit of thirty milli Sievert (30 mSv).

## SCHEDULE IX

[See Rule 18(3)]

LEAKAGE AND SAFETY RELATED TESTS

1. **Wet storage irradiators**– (1) The resin bed in the water conditioning system shall be daily checked with a radiation survey instrument. The survey instrument should be sensitive enough to detect minimum radioactivity of 2000 Bq. in the resin bed. In the event of detection of activity above this level, water circulation system shall be stopped and irradiator shall be withdrawn from service.

   (2) In addition the pool water shall be checked for contamination by using an on-line radiation monitor on a pool water circulating system.

   (3) The detection of above normal radiation levels must activate an alarm. Activation of alarm must automatically cause the water purification system to shut off.

1. **Dry storage irradiators**– Swipe test using a moist paper of 100 square centimeter area shall be conducted weekly on the closest accessible surface near the source in storage condition. Activity on the sample shall be counted by a radiation survey

instrument with a minimum detection capability of 2000 Bq on swipe sample. In the event of detection of activity above this level, irradiator shall be withdrawn from service.

## FORM I

[See rule 4(1)]

APPLICATION FOR OBTAINING A LICENSE FOR AN IRRADIATION FACILITY

1. Name of the applicant
2. Address of the applicant
3. Installation for which approval is applied for
4. Name and designation of the Head of Installation
5. Names, qualifications and experience of Personnel

| *Category Personnel* | *Name* | *Academic Qualification* | *Type of Training experience* | *When and where trained* | *Duration of training* |
|---|---|---|---|---|---|
| Operator | | | | | |
| Radiological Safety | | | | | |
| Quality Control Officer | | | | | |

6. Proposed date of starting the irradiation facility:
7. Details about the Irradiation Facility :
   (1) Identification number of the facility
   (2) Location and address
   (3) Source details
   Name of radionuclides — Activity
   Radiation generating plant — Energy — Output
   X-ray unit
   Electron Accelerator
   (4) Name of the supplier and his address
   (5) Purpose for which the irradiation facility will be used
8. Please attach the following additional information to the application :

(1) A site plan (1:500 scale or as appropriate) of the installation indicating the location of buildings including residential complexes.

(2) Architectural blue prints (appropriate scale) showing layout of equipments.

(3) Complete design drawing of the facility

(4) Description of the organizational structure including delegation of authority and responsibility for operation at the facility.

9. Please indicate as appropriate:

(1) the irradiation facility is yet to be built

(2) the irradiation facility is already built and equipped

(3) Existing irradiation facility is to be modified as per details enclosed.

10. I hereby certify that :

(1) all the statements made above are correct to the best of my knowledge and belief.

(2) no operations will be carried out for purposes other than those specified under item 7 (5) of this form.

(3) all provisions of the Atomic Energy (Control of Irradiation of Food) Rules, 1996 shall be strictly complied with.

(4) the irradiation facility shall not be transferred/sold/rented by me/us to another without the prior permission of the competent authority.

(5) no radiation source for the irradiation unit will be transported without the prior permission of the competent authority.

(6) full facilities will be accorded by me/us to any authorised representative of the competent authority or the licensing authority to inspect the installation at any time.

(7) radiation surveillance and medical surveillance of all persons engaged in radiation work as required by the competent authority will be duly carried out at my/our expense.

(8) all recommendations that may be made from time to time by the competent authority in respect of radiation safety measures will be duly implemented.

(9) duly qualified/experienced radiological safety officers/ quality officers will be appointed before the commencement of the operation of the facility.

(10) any changes in the personnel listed in this application will be intimated forthwith to the licensing authority and the competent authority.

(11) The rules regarding decommissioning, disposal of decayed sources and reuse of the site of the decommissioned facility shall be strictly complied with.

Date : Signature of the Applicant Institution and Seal

**FORM II**

[See rule 5 (2)]

CERTIFICATE OF APPROVAL FOR IRRADIATION FACILITY

Mr./Messers.........................of..............having complied with the conditions prescribed in the Atomic Energy (Control of Irradiation of Food) Rules, 1996 for the approval of the irradiation facility is/are hereby granted the certificate of approval for his/their irradiation facility.

The description of the irradiation facility and other details are shown as under :

1. Identification Number
2. Name and Addresses
   - (i) Licensee
   - (ii) Radiological Safety Officer
   - (iii) Quality Control officer
   - (iv) Operator
3. Kind and nature of the food product to be irradiated
4. Radiation source used
   - (i) Type of Source
   - (ii) Total activity content in the case of radionuclide source
   - (iii) Type of operation–Batch or Continuous
   - (iv) Source product geometry
   - (v) Dose rate

5. Specific operation conditions

COMPETENT AUTHORITY
(Seal of Office)

## FORM III

[See rule 5 (1)]

### APPLICATION FOR OBTAINING THE CERTIFICATE OF APPROVAL FOR AN IRRADIATION FACILITY

1. Name of the applicant
2. Address of the applicant
3. Installation for which approval is applied for
4. Names, qualifications and experience of Personnel

| *Category Personnel* | *Name* | *Academic Qualification* | *Type of Training experience* | *When and where trained* | *Duration of training* |
|---|---|---|---|---|---|
| Operator | | | | | |
| Radiological Safety Officer | | | | | |
| Quality Control Officer | | | | | |

5. Proposed date of starting the irradiation facility
6. Are the personnel provided with facilities for:
   (a) Personnel dose monitoring
   (b) Medical Surveillance
7. Details about the Irradiation Facility:
   (1) Identification number of the facility
   (2) Location and its address
   (3) Source detail
   Name of radionuclides
   Radiation generating plant — Activity, Energy, output
   X-ray unit
   Electron Accelerator

(4) Name of the supplier and his address

(5) Purpose for which the irradiation facility will be used.

8. Please furnish the following :

(1) A site plan (1:500 scale or as appropriate) of the installation indicating the location of buildings including residential complexes. Occupancy within 50 meters radius of the facility.

(2) Architectural blue prints (appropriate scale) showing layout of equipment.

(3) Details on geology of the location, water table, soil characteristics, seismicity.

(4) Complete design drawing of the facility including details of shielding and surrounding of source, wall thickness and labyrinth access if applicable; openings, voids, reinforcements, mechanical and electrical safety systems, ventilation, fire protection systems.

(5) Source movement system (where appropriate).

(6) Safety analysis report to demonstrate the adequacy of radiation safety under normal and anticipated accident conditions as detailed in rule 20 and 21.

(7) Operating and emergency procedures.

(8) List of calibrated radiation monitoring equipment in working condition.

(9) Description of the organizational structure including delegation of authority and responsibility for operation of the facility.

9. Any other information which the competent authority may deem necessary to assess the safety status of the irradiation facility.

10. Please indicate as appropriate:

(a) the irradiation facility is yet to be built

(b) the irradiation facility is already built and equipped

(c) existing irradiation facility is to be modified as per details enclosed.

11. I hereby certify that:
    (1) all the statements made above are correct to the best of my knowledge and belief.
    (2) no operations will be carried out for purposes other than those specified under item 7(5) of this form.
    (3) all provisions of the Atomic Energy (Control of Irradiation of Food) Rules, 1996, shall be strictly complied with.
    (4) the irradiation facility shall not be transferred/sold/rented by me/us to another without the prior permission of the competent authority.
    (5) no radiation source for the irradiation unit will be transported without the prior permission of the competent authority.
    (6) full facilities will be accorded by me/us to any authorized representatives of the competent authority or the licensing authority to inspect the installations at any time.
    (7) radiation surveillance and medical surveillance of all persons engaged in radiation work as required by the competent authority will be duly carried out at my/our expense.
    (8) all recommendations that may be made from time to time by the competent authority in respect of radiation safety measures will be duly implemented.
    (9) duly qualified/experienced radiological safety officers/operators/quality officers will be appointed before the commencement of the operation of the facility.
    (10) the rules regarding decommissioning disposal of decayed sources and reuse of the site of the decommissioned facility shall be strictly complied with.

Date:

Signature of the Applicant
Seal of the Institution

## FORM IV

[See rule 4(2)]

LICENCE FOR OPERATING AN IRRADIATION FACILITY

Mr./Messers..................of..................having undertaken to comply with all the conditions prescribed in the Atomic Energy (Control of Irradiation of Food) Rules, 1996 after having paid the prescribed licence fee and having obtained the certificate of approval from the competent authority is/are hereby authorised to commission and operate the irradiation facility described in the application for licence.

This licence is issued on..................and shall be valid upto ..........subject to the conditions printed overleaf.

### Conditions of Licence

1. This licence may be suspended or cancelled if any declaration made or information given in the application is found to be false or if any undertaking given in such application is not carried out.
2. No operations will be carried out for purposes other than those specified under item 7(5) of the application for the licence and item 7(5) of the application for the certificate of approval.
3. All provisions of the Atomic Energy (Control of Irradiation of Food) Rules, 1996, shall be strictly complied with.
4. The irradiation facility shall not be transferred/sold /rented by me/us to another without the prior permission of the competent authority.
5. No radiation source for the irradiation unit will be transported without the prior permission of the competent authority.
6. Full facilities will be accorded by me/ us to any authorized representatives of the competent authority or the licensing authority to inspect the installation at any time.

7. Radiation surveillance and medical surveillance of all persons engaged in radiation work as required by the competent authority will be duly carried out at my/our expense.
8. All recommendations that may be made from time to time by the competent authority in respect of radiation safety measures will be duly carried out at my/our expense.
9. Duly qualified/experienced radiological safety officers/ operators/quality officers will be appointed before the commencement of the operation of the facility.
10. Any changes in the personnel listed in this application will be intimated forthwith to the licensing authority and the competent authority.
11. The rules regarding decommissioning, disposal of decayed sources and reuse of the site of the decommissioned facility shall be strictly complied with.
12. A licence shall be valid for a *period of three years* from the date of issue of the licence.

## FORM V

[See rule 11 (1) and (4)]

RECORD OF FOOD IRRADIATION

1. Identification number of the irradiation facility
2. Name and address of the operator and Quality Control Officer
3. Serial number of the batch
4. Date of irradiation
5. Type of source (radionuclide source or X-ray or electron machine source.
6. Irradiation Dose
7. Type of Dosimeters used and their calibration
8. Minimum and Maximum limits of radiation dose
9. Nature, quality and quantity of food to be irradiated
10. Purpose of irradiation
11. Type of packaging used during radiation treatment, if any

12. *Specific irradiation conditions used
13. Any other incidents and anomalies observed during irradiation treatment.

* Information necessary are : Commissioning details, dose mapping for individual products, source position data, product loading pattern, conveyor operation number, duration of process interruption, starting and stopping time of processing and irradiation time.

## Prevention of Adulteration Rules (1955)

### [PART XVII][1]

### IRRADIATION OF FOOD

[2][73. For the purpose of this chapter unless the context otherwise requires:

(a) 'Irradiation' means any physical procedure, involving the intentional exposure of food to ionizing radiations.

(b) 'Irradiation facility' means any facility which is capable of being utilized for treatment of food by irradiation.

(c) 'Operator who satisfies the qualifications and requirements as for training specified in Schedule II of the Atomic Energy (Control of Irradiation of Food) Rules, 1991.

(d) 'Irradiated food' means articles of food subjected to radiation by:

  (i) Gamma rays;

  (ii) X-rays generated from machine sources operated at or below an energy level of 5 million electron volts; and

  (iii) Sub-atomic particles, namely, electrons generated from machine sources operated at or below an energy level of 10 million electron volts, to dose levels as specified in Schedule I of the Atomic Energy (Control of Irradiation of Food) Rules, 1991.

---

1. Ins. by GSR 614 (E) dt. 9.8.1994 (w.e.f. 9.8.1994) as corrected by GSR 60 (E) dt. 7.2.1995.
2. –do–

74. **Dose of Irradiation.**–(1) Save as provided in sub-rule (2), no food shall be irradiated.

(2) No article of food permitted for irradiation specified in column 2 of the Table given below shall receive the dose

| *Sl. No.* | *Name of Foods* | *Dose of Irradiation (kGy)* | | |
|---|---|---|---|---|
| | | *Minimum* | *Maximum* | *Overall Average* |
| 1. | Onions | 0.03 | 0.09 | 0.06 |
| 2. | Spices | 6 | 14 | 10 |
| 3. | Potatoes | 0.06 | 0.15 | 0.10 |
| 4. | Rice | 0.25 | 1.0 | 0.62 |
| 5. | Semolina (Sooji or Rawa), Wheat atta and Maida | 0 .25 | 1.0 | 0.62 |
| 6. | Mango | 0.25 | 0.75 | 0.50 |
| 7. | Raisins, Figs and Dried Dates | 0.25 | 0.75 | 0.50 |
| 8. | Ginger, Garlic and Shallots (Small Onions) | 0.03 | 0.15 | 0.09 |
| 9. | Meat and Meat Products including Chicken | 2.5 | 4.0 | 3.25 |
| 10. | Fresh Sea foods | 1.0 | 3.0 | 2.00 |
| 11. | Frozen Sea foods | 4.0 | 6.0 | 5.00 |
| 12. | Dried Sea foods | 0.25 | 1.0 | 0.62 |
| 13. | Pulses | 0.25 | 1.0 | 0.62 |

of irradiation in excess of the quantity specified in column 3 of the said Table at the time of irradiation:–

(3) Routine quantitative dosimetry shall be made during operation and record kept of such measurement as provided under Department of Atomic Energy (Control of Irradiation of Food) Rules, 1991.

75. **Requirement for the Process of Irradiation.**–
*(1) Approval of facilities.*– No irradiation facility shall be used for the treatment of food unless such facility:–

(a) has been approved and licensed under the Atomic Energy (Control of Irradiation of Food) Rules, 1991;

(b) complies with the conditions for approval, operation, licence and process control prescribed under the Atomic Energy (Control of Irradiation of Food) Rules, 1991;

(c) carries out irradiation in accordance with the provisions of the Atomic Energy (Control of Irradiation of Food) Rules, 1991.

*(2)* Foods once irradiated shall not be re-irradiated unless specifically so permitted by the Licensing Authority for the irradiation process control purposes.

*(3)* No food/irradiated food shall leave the irradiation facility unless it has been irradiated in accordance with the provisions of Atomic Energy (Control of Irradiation of Food) Rules, 1991 and a certificate of irradiation indicating the dose of irradiation and the purpose of irradiation is provided by the competent authority.

76. **Restrictions on Irradiation of Food.**–(a) The irradiation shall conform to the dose limit and the radiation source to the specific conditions prescribed for each type or category of food specified for treatment by irradiation, under the Atomic Energy (Control of Irradiation of Food) Rules, 1991.

(b) Food which has been treated by irradiation shall be identified in such a way as to prevent its being subjected to re-irradiation.

(c) The irradiation shall be carried out only by personnel having the minimum qualifications and training as prescribed for the purpose under the Atomic Energy (Control of Irradiation of Food) Rules, 1991.

(d) Food once irradiated shall not be re-irradiated unless specifically so permitted under these rules.

**77. Record of Irradiation of Food.**–Any treatment of food by irradiation shall be recorded by an officer authorised by the competent authority as specified under the Atomic Energy (Control of Irradiation of Food) Rules, 1991 as follows;–

(i) Name of the article:

(ii) Licence No.:

(iii) Name, address and other details of Licensee:

(iv) Purpose of Irradiation:

(v) Source of Irradiation:

(vi) Date of Irradiation:

(vii) Dose of Irradiation:

(viii) Serial Number of Batch:

(ix) The nature, quality of food to be irradiated and the Batch number:

(x) Quantity of food irradiated:

(xi) Physical appearance of article: before and after irradiation:

(xii) Type of packaging used during the irradiation treatment and for packing the irradiated food.

**78. Standards of Irradiated food.**—The irradiated foods shall comply with all the provisions of Prevention of Food Adulteration Act, 1954, and rules made thereunder specifying standards of such food.

# INDEX